AF717565

Nuclear and Particle Physics

Nuclear and Particle Physics

Suresh Chandra
Mohit K. Sharma

Alpha Science International Ltd.
Oxford, U.K.

Suresh Chandra
Head
Department of Physics
Lovely School of Sciences
Lovely Professonal University
Phagwara, Punjab

Mohit K. Sharma
III-A/21
Vaishali, Gaziabad
Uttar Pradesh

Copyright © 2012

ALPHA SCIENCE INTERNATIONAL LTD.
7200 The Quorum, Oxford Business Park North
Garsington Road, Oxford OX4 2JZ, U.K.

www.alphasci.com

All rights reserved. No part of this publication may be reproduced, stored in a retrieval system, or transmitted in any form or by any means, electronic, mechanical, photocopying, recording or otherwise, without prior written permission of the publisher.

Printed from the camera-ready copy provided by the Authors.

ISBN 978-1-84265-745-4

Printed in India

Preface

Nuclear & Particle Physics is one of the fundamental courses for graduate and postgraduate students of Physics as well as Chemistry. A number of books have been written on this topics and every one claims his/her presentation to be lucid and student friendly. However, this topic has always been little difficult for a common student. We also made this attempt to present the topic in a student friendly manner using simple language. In the presentation, we tried to give exercises which help in order to make the topics as simple as possible. We hope, this attempt would be appreciated by the students and teachers. We shall appreciate receiving comments/suggestions, which can be sent at the following emails also:

suresh492000@yahoo.co.in mohit781987@gmail.com

While preparing the manuscript, we have been helped, advised, and encouraged by our seniors, friends, and colleagues working in various institutions in India as well as abroad, and by our friends in personal life. We are heartily thankful to all of them. We are grateful to Prof. J.V. Narlikar, Prof. Dr. W.H. Kegel, Prof. S.P. Khare, for their encouragement. For preparation of the manuscript, we have been helped by Mr. B.K. Kumthekar, Mr. G.M. Dak and Mr. Amit Kumar. We are thankful to them. Special thanks are due to Mrs. Purnima Sharma for her valuable cooperation for sharing all responsibilities of family affairs, so that we could spend our time for this book. We are deeply grateful to our family members who always have been a source of encouragement and happiness for us. In the last but not the least, we are highly thankful to M/S Narosa Publishing House Pvt. Ltd., New Delhi for publishing this book.

Suresh Chandra

Mohit K. Sharma

Contents

I. Introduction

We are aware of three states of matter, commonly known as the solid, liquid and gas. The basic constituents of a matter are molecules, which are formed due to combination of various atoms. According to the Dalton hypothesis, proposed in 1808, an atom was the smallest unit. That is the atom was indivisible. The atoms of different elements were considered to be completely different in nature from one another. With the discovery of electron, a particle having negative charge (by convention), by J.J. Thomson in 1897, the idea of indivisibility of atom was shattered down. It was therefore concluded that an atom was consisted of various kinds of particles. Rutherford experiment on scattering of α-particles led to the concept of positively charged nucleus in an atom. In this chapter, we shall discuss about the discovery of nucleus of an atom and some parameters about it.

1. Discovery of nucleus

With the discovery of electron, the idea of indivisibility of atom was shattered down. Since an atom as a whole has no net charge (*i.e.*, it is neutral), it must contain equal amount of negative and positive charges. Also the mass of electron was found to be negligible in comparison to that of the lightest atom (hydrogen atom). Therefore, it was concluded that the mass of an atom is due to the mass of positively charged matter in the atom.

Rutherford and his coworkers performed an experiment on the scattering of α-particles by a thin foil of matter. In this experiment, some α-particles were found scattered through an angle greater than 90°. (This is known as the back-scattering.) In order to explain the back-scattering of α-particles, Rutherford in 1911 proposed a nuclear model of atom. According to this model, all the positive charge and hence most of the mass of the atom was confined to a very small volume, called the nucleus, situated at the center of the atom. It was presumed that a nucleus had positively charged particles, named protons. Hence, the nucleus of an atom was discovered in 1911 as a consequence of Rutherford scattering of α-particles. At that time, obviously, the known elementary particles were the electron and proton.

1.1 Proton-electron theory

It was found that all the atomic masses were integer multiples of the mass of hydrogen atom. The slight change from the whole number was assigned to the presence of two or more isotopes of the element. This hydrogen theory confirms the presence of protons inside the atom. Since electrons (β-particles) are known to be emitted from some nuclei, it was natural to believe in the beginning that the electrons were the constituents of all nuclei. To account for the mass number A (which is an integer), it was necessary to assume that the nucleus would be consists of A protons. But, if it were the case, the charge on the nucleus would be positive A, which is not equal to the atomic number (number of electrons) Z. In order to short out this controversy, it was presumed that in addition to A protons the atomic nuclei would contain $(A - Z)$ electrons, so that the net charge on the nucleus is positive Z. These electrons would contribute a negligible amount to mass of the nucleus, but would make the charge positive Z, as required.

This theory could explain the emission of α and β-particles. Presence of electron in the nucleus showed that under the appropriate conditions, one of the electrons might be ejected (emission of a β-particle). For emission of an α-particle, two electrons could combine with four protons before emitting. On the other hand, this theory lead to a number of contradictions with experiments. Some of them are as the following.

(i) Finite size

As the electrons are spherical charges of finite dimensions[1] with radius 2.818×10^{-15} m, therefore, for a heavy nucleus having large number of electrons, it is difficult to accommodate them in it.

---

[1]Classical electron radius

$$r_e = \frac{1}{4\pi\epsilon_0}\frac{e^2}{m_e c^2} = 2.818 \times 10^{-15} \text{ m}$$

proton radius

$$r_p = \frac{1}{4\pi\epsilon_0}\frac{e^2}{m_p c^2} = 1.535 \times 10^{-18} \text{ m}$$

neutron radius

$$r_n = \frac{1}{4\pi\epsilon_0}\frac{e^2}{m_n c^2} = 1.533 \times 10^{-18} \text{ m}$$

(ii) Spin consideration

As each of the electron and proton has spin 1/2, nucleus must have an integral spin when it has an even number of particles (protons + electrons) and a half integral spin when it has an odd number of particles. But, experiments show that the spin depends only on the mass number A of the nucleus, and has integer value when A is even and half integer value when A is odd. According to the proton-electron theory, for an atom ${}_Z\mathrm{X}^A$, the number of particles inside the nucleus will be A + (A – Z) = 2 A – Z (A protons and (A – Z) electrons). For example, for ${}_7\mathrm{N}^{14}$, the nucleus should have 21 particles and therefore the spin should be an half integer, which is against the observed facts.

(iii) Statistics

From the band spectra and Raman spectra the statistics of ${}_7N^{14}$ was found to be the Bose-Einstein statistics. If the nucleus was made of $(2A - Z) = 23$ fermions (protons and electrons), they should obey the Fermi-Dirac statistics.

(iv) Magnetic moment consideration

Since the nucleus has charged particles and the rotation of the charged particles gives rise to a magnetic moment. The magnetic moment $(eh/4\pi m)$ of a proton is nearly 1837 times less than that of an electron. Total magnetic moment of all protons is small in comparison with the magnetic moment of an electron. The value of magnetic moment shows that the electron is not constituent of a nucleus.

(v) Wave mechanical consideration

On the basis of uncertainty principle, it was found that electrons cannot exist in a nucleus. It can be verified in following manner.

Let us imagine that an electron exists in a nucleus. The diameter of the nucleus is of the order of $\sim 10^{-14}$ m. Thus, the maximum uncertainty in the position of the electron can be equal to the diameter of the nucleus. According to the Heisenberg's Uncertainty Principle, the minimum uncertainty in the momentum of the electron should be

$$\Delta p = \frac{h/2\pi}{\Delta x} = \frac{6.626 \times 10^{-34}}{2 \times 3.14 \times 1 \times 10^{-14}} = 1.05 \times 10^{-20} \text{ kg-m}$$

The momentum of electron may be taken of the same order of magnitude as the uncertainty in its momentum (*i.e.*, $p \approx 10^{-20}$ kg-m). So much momentum of electron can only be possible when its velocity is comparable to the speed of light. Therefore, the total energy E of the electron is expressed through the relativistic formula

$$E^2 = p^2c^2 + m_0^2c^4$$

where m_0 is the rest-mass of electron. For relativistic electrons, we have $p^2c^2 >> m_0^2c^4$, and therefore,

$$E^2 = p^2c^2 \qquad E = pc \qquad \Delta E = c\Delta p$$

Hence, the minimum uncertainty in the energy of the electron to exist inside the nucleus is

$$\Delta E = 3 \times 10^8 \times 1.05 \times 10^{-20} = 3.16 \times 10^{-12} \text{ J}$$
$$= \frac{3.16 \times 10^{-12}}{1.602 \times 10^{-19}} = 2 \times 10^7 \text{ eV} = 20 \text{ MeV}$$

Thus, for the electron to exist in the nucleus, the minimum uncertainty in the energy is about 20 MeV. It is several orders larger than the energy of β-particle emitted from the nucleus. Since the uncertainty in a quantity cannot be larger than the quantity itself, it shows that electron cannot exist in the nucleus.

(vi) Compton wavelength

Several theoretical considerations suggest that a bound fundamental particle cannot be localized in the region smaller than its Compton wavelength h/m_0c.

$$\lambda = \frac{h}{m_0c} = \frac{6.626 \times 10^{-34}}{9.1 \times 10^{-31} \times 3 \times 10^8}$$

$$= 243 \times 10^{-14} \approx 250 \text{ (Nuclear diameter)}$$

This excludes the possibility of keeping the electrons inside the nucleus.

1.2 Proton-neutron theory

Based on the experimental discovery of neutron, Heisenberg in 1932 suggested that the nuclei might be composed of protons and neutrons rather than of protons and electrons. Thus, for an atom $_Z\text{X}^A$, the nucleus contains Z protons and $(A - Z)$ neutrons. The total number of particles inside the nucleus is A. The nucleus is surrounded by Z electrons to make the atom electrically neutral. This theory avoids the failure of the proton-electron theory as the following:

(i) Finite size

Since the mass of a neutron is nearly equal to that of a proton, the sizes of both are nearly equal. Thus, it is not difficult to accommodate protons and neutrons in a nucleus.

(ii) Spin consideration

Both the proton and neutron have the spin quantum number 1/2. Hence, according to the quantum theory, the resultant spin of A nucleons will be an integer when A is an even integer or a half integer when A is an odd integer. This is in agreement with all the experimental observations.

(iii) Statistics

Since the neutrons and protons both are fermions with spin quantum number 1/2, the statistics based on the resultant spin of nucleons agrees with the experimental observations.

(iv) Magnetic moment consideration

The magnetic moments of proton and neutron both are almost the same. The values of total magnetic moment of various nuclei are in agreement with the observations.

(v) Wave mechanical consideration

Also, in case of proton as well as neutron, the uncertainty in position is of the order of $\sim 10^{-14}$ m (*i.e.*, radius of the nucleus). Thus, the uncertainty in the momentum is

$$\Delta p = \frac{h/2\pi}{\Delta x} = \frac{6.626 \times 10^{-34}}{2 \times 3.14 \times 1 \times 10^{-14}} = 1.05 \times 10^{-20} \text{ kg-m}$$

The momentum of proton as well as neutron may be taken of the same order of magnitude as the uncertainty in its momentum (*i.e.*, $p \approx 10^{-20}$ kg-m). Total energy E of a proton or neutron is expressed as

$$E^2 = p^2c^2 + m_0^2c^4$$

$$= (10^{-20})^2 \times (3 \times 10^8)^2 + (1.67 \times 10^{-27})^2(3 \times 10^8)^4$$

$$= 9 \times 10^{-24} + 2.25 \times 10^{-20} = 2.25 \times 10^{-20}$$

so that the energy is

$$E = 1.5 \times 10^{-10}\ \mathrm{J} = \frac{1.5 \times 10^{-10}}{1.602 \times 10^{-19} \times 10^{6}} = 936\ \mathrm{MeV}$$

This value is slightly greater than the rest energy of a proton or a neutron. Hence, the kinetic energy of a neutron or proton in the nucleus is of the order of few MeV and it may be possible for a free neutron or proton to exist in the nucleus.

(vi) Compton wavelength

Compton wavelength of neutron is given by the relation.

$$\lambda = \frac{h}{m_0 c} = \frac{6.626 \times 10^{-34}}{1.67 \times 10^{-27} \times 3 \times 10^{8}}$$

$$= 0.13 \times 10^{-14}\ \mathrm{m} = 0.13\ \text{(Nuclear diameter)}$$

Thus, the protons and neutrons can be squeezed into the nuclear volume.

(vii) β-decay

The β-ray emission from some nuclei can be suggested as the following. The electron does not pre-exist in the nucleus but is formed at the time of emission as indicated by the equation

$$n \rightarrow p + e^{-} + \overline{\nu}$$

β^{+}-decay is due to the reaction

$$p \rightarrow n + e^{+} + \nu$$

1.3 Neutron-positron theory

According to this theory, a nucleus consists of neutrons and positrons. Since the mass of positron (β^{+}) is equal to that of electron, this theory met to the same difficulties with which proton-electron theory came across.

2. Nomenclature

An individual atomic species is called a nuclide. A nuclide X having Z protons and A nucleons is represented as ${}_Z\mathrm{X}^A$, ${}^A\mathrm{X}_Z$ or ${}^A_Z\mathrm{X}$. Nuclides or atoms having the same atomic number (Z) are called the **isotopes**, whereas the atoms with the same number of neutrons ($N = A - Z$) are called the **isotones**. Atoms having the same mass number A are called the **isobars**. Nuclide having the same neutron excess ($N - Z = A - 2Z$) or asymmetry in neutron - proton number are called ionospheres. **Isomers** are metastable excited levels with a fairly long life-time.

Table 1.1: Nomenclature of nuclear species

Nomenclature	Characteristics	Examples
Nuclides	Z, A	${}_{8}O^{16}$, ${}_{50}Sn^{120}$, ${}_{92}U^{238}$
Isotopes	constant Z	${}_{7}N^{13}$, ${}_{7}N^{14}$, ${}_{7}N^{15}$
Isotones	constant $N = A - Z$	${}_{6}C^{14}$, ${}_{7}N^{15}$, ${}_{8}O^{16}$
Isobars	constant A	${}_{6}C^{14}$, ${}_{7}N^{14}$
Isodiapheres	constant $N - Z = A - 2Z$	(${}_{6}C^{12}$, ${}_{7}N^{14}$, ${}_{8}O^{16}$)
		(${}_{88}Ra^{226}$, ${}_{86}Rn^{222}$)
Isomers	constant Z and A	${}_{35}Br^{80m}$(4.4 hr),
		${}_{35}Br^{80}$(18 min)

2.1 Mirror nuclei

A pair of nuclei with the same mass number A but with the proton number Z and neutron number N interchanged, are known as the mirror nuclei. Some pairs of mirror nuclei are as the following:

$$\left({}^{13}N_7,\ {}^{13}C_6\right),\ \left({}^{23}Mg_{12},\ {}^{23}Na_{11}\right),\ \left({}^{25}Al_{13},\ {}^{25}Mg_{12}\right)$$

Exercise 1: Calculate the energy equivalent to one amu, mass of electron, mass of proton and mass of neutron. Given, 1 amu = 1.660×10^{-27} kg, mass of electron $m_e = 9.109 \times 10^{-31}$ kg, mass of proton $m_p = 1.673 \times 10^{-27}$ kg, mass of neutron $m_n = 1.675 \times 10^{-27}$ kg and speed of light $c = 2.998 \times 10^8$ m/s.

Solution: The energy E, equivalent to 1 amu is

$$\begin{aligned} E = mc^2 &= 1.660 \times 10^{-27} \times (2.998 \times 10^8)^2 = 1.4920 \times 10^{-10} \text{ J} \\ &= \frac{1.4920 \times 10^{-10}}{1.602 \times 10^{-19}} = 931 \times 10^6 \text{ eV} = 931.3 \text{ MeV} \end{aligned}$$

Energy equivalent to the mass of one electron is

$$E_e = \frac{9.109 \times 10^{-31} \times 931.3}{1.660 \times 10^{-27}} = 0.5105 \text{ MeV}$$

Energy equivalent to the mass of one proton is

$$E_p = \frac{1.673 \times 10^{-27} \times 931.3}{1.660 \times 10^{-27}} = 938.59 \text{ MeV}$$

Energy equivalent to the mass of one neutron is

$$E_n = \frac{1.675 \times 10^{-27} \times 931.3}{1.660 \times 10^{-27}} = 939.79 \text{ MeV}$$

3. Parameters of a nucleus

Size of a nucleus is very small as compared to that of the lightest atom (hydrogen atom). The nuclear diameter is of the order of 10^{-14} m whereas the atomic diameter is of the order of 10^{-10} m. Neutrons and protons are the main constituents of a nucleus. Mass of a neutron is nearly equal to that of a proton, which is about 1840 times more than that of an electron. Neutrons and protons are collectively known as the nucleons. The shape of the nucleus is taken as spherical because for a given volume the sphere possesses the least surface area and will therefore provide the maximum short range binding forces between the nucleons (protons and neutrons) in the nucleus.

3.1 Size

The nucleus is found to be incompressible and there is sufficient evidence that nuclear density ρ remains approximately constants over most of the nuclear volume and then decreases rapidly to zero. This means that the nuclear volume is approximately proportional to number of nucleons, *i.e.* mass number A.

$$\text{volume} \propto A \qquad \text{or} \qquad R = R_0 \, A^{1/3}$$

where R is the radius of the nucleus and R_0 a constant whose value is about 1.2 $\times 10^{-15}$ m.

For two nuclei with atomic masses A_1, A_2 and radii R_1, R_2, respectively have the relation

$$\frac{R_1}{R_2} = \left(\frac{A_1}{A_2}\right)^{1/3}$$

Exercise 2: Two elements X and Y have atomic numbers 4 and 32, respectively. Calculate the ratio of their nuclei.

Solution: The radius R of a nucleus is expressed as

$$R = R_0 \ A^{1/3}$$

where R_0 is a constant and A the atomic number. For the nuclei X and Y, the radii are

$$R_X = R_0 \ 4^{1/3} \qquad \text{and} \qquad R_Y = R_0 \ (32)^{1/3}$$

Thus, we have $R_Y/R_X = 2$.

3.2 Density

Now, we have

$$\text{Mass of nucleus} = A \text{ amu} = A \times 1.66 \times 10^{-27} \text{ kg}$$

$$\text{Volume of nucleus} = \frac{4}{3} \pi R^3 = \frac{4}{3} \pi R_0^3 \ A$$

$$\text{Density of nucleus} = \frac{\text{mass}}{\text{volume}} = \frac{A \times 1.66 \times 10^{-27}}{4\pi R_0^3 \ A/3}$$

$$= \frac{1.66 \times 10^{-27} \times 3}{4\pi R_0^3} \approx 2 \times 10^{17} \text{ kg/m}^3$$

where we have used $R_0 = 1.2 \times 10^{-15}$ m. Hence, the density of a nucleus is enormously high.

3.3 Charge distribution

The experiments have been performed and analyzed for several nuclei at several incident electron energies. All the results can be approximately explained by a charge distribution

$$\rho(r) = \frac{\rho_0}{1 + \exp[(r - R)/a]} \tag{1.1}$$

A plot of the variation of $\rho(r)$ versus r is shown in Figure 1. It is found that the results for all nuclei are reasonably well approximated by equation (1.1) with

$$\rho_0 \approx 1.65 \times 10^{44} \ \text{nucleons/m}^3 = 0.165 \ \text{nucleons/fm}^3$$

$$R \approx 1.07 \ A^{1/3} \ \text{fm}$$

$$a \approx 0.55 \ \text{fm}$$

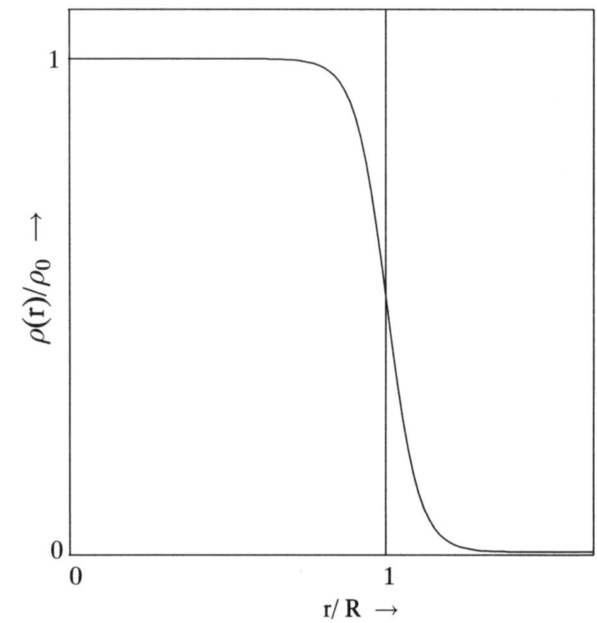

Figure 1: Variation of charge distribution $\rho(r)/\rho_0$ as a function of r/R. For the graph, we have used $R = 3.0$ and $a = 0.15$.

3.4 Parity of a nucleus

Wave function of a nucleus describes its behaviour as a function of time and space. The behaviour of the wave function $\Psi(\vec{r})$ under reflection of space coordinates ($\vec{r}$ changing to $-\vec{r}$) determines about the parity of the nucleus. Under the space reflection, sign of the wave function may change or may remain unchanged. When the sign changes, the nucleus is said to have an odd parity and when the sign remains unchanged, the nucleus is said to have an even parity. It can be expressed as

$$\Psi(-\vec{r}) = -\Psi(\vec{r}) \qquad \text{odd parity}$$

$$\Psi(-\vec{r}) = \Psi(\vec{r}) \qquad \text{even parity}$$

The product of two odd wave functions results in an even parity wave function whereas the product of an even wave function and an odd wave function results in an odd parity wave function.

Parity of an isolated proton or neutron wave function is even while that of a π meson is odd. For a system of interacting nucleons, the parity, denoted by π is expressed as

$$\pi = (-1)^l$$

where l is the orbital quantum number of the system.

The law of conservation of parity states that the parity of a system of nuclear particles remains constant under any reaction or transformation. As a consequence of conservation of parity, a nucleus or a system of nuclear particles cannot co-exist in the states of different parities.

3.5 Atomic mass unit

John Dalton was the first to suggest that the mass of one atom of hydrogen be used as the atomic mass unit. Francis Aston, inventor of mass spectrometer, later on used 1/16 (one-sixteenth) of the mass of one atom of oxygen-16 ($_8O^{16}$) as his unit for atomic mass. Before 1961, the atomic mass unit was thus defined as 1/16 (one-sixteenth) of the mass of one atom of oxygen-16 ($_8O^{16}$). Since 1961, the atomic mass unit is defined as 1/12 (one-twelfth) of the mass of one atom of carbon-12 ($_6C^{12}$). The value of the atomic mass unit is as the following.

$$1 \text{ amu} = 1.66053878 \times 10^{-27} \text{ kg}$$

Using the energy-mass relation, equivalent of one atomic mass unit is 931.49 MeV. The mass of some common particles and their energy equivalent are as the following.

$$\text{Mass of electron} = 0.000549 \text{ amu} = 0.511 \text{ MeV}$$

$$\text{Mass of proton} = 1.007276 \text{ amu} = 938.27 \text{ MeV}$$

$$\text{Mass of neutron} = 1.008665 \text{ amu} = 939.56 \text{ MeV}$$

Atomic masses of various nuclides are given in the appendix.

4. Energies of nucleons in a nucleus

We shall often discuss about the kinetic energies of nucleons. Let us here estimate an order-of-magnitude of the energies of nucleons. The energies of β and γ rays emitted from nuclei are typically of the order of 1 MeV. These rays are however emitted during the transitions of nucleons from one state to another and therefore the energies of β and γ rays are the differences between the energies in two different states. The actual nucleon energy should be much larger than those of β and γ rays. One approach to estimate energies of nucleons may be to calculate the electrostatic energy E_c required to insert a proton into a nucleus. This is approximately

$$E_c = \frac{Ze^2}{4\pi\epsilon_0 R}$$

For a medium weight nucleus with $Z = 50$ and $A = 120$, we have

$$E_c = \frac{9 \times 10^9 \times 50 \times (1.602 \times 10^{-19})^2}{1.07 \times 10^{-15} \times 120^{1/3}} = 2.19 \times 10^{-12} \text{ J}$$

$$= \frac{2.19 \times 10^{-12}}{1.602 \times 10^{-19}} = 13.7 \times 10^6 \text{ eV} = 13.7 \text{ MeV}$$

Thus, 13.7 MeV energy would be released if a proton is allowed to enter in the nucleus. One may guess that energies of nucleons in the nucleus are of the order of 10 MeV. We shall later on find that this is an underestimate, but its order of magnitude is correct. As the velocity of a 10 MeV nucleon is only 15% of the speed of light, it shows that the relativistic effects are not important for the motion of nucleons in a nucleus. Hence, the non-relativistic relations between mass, velocity, momentum and kinetic energy may be used freely.

5. Nucleus is a quantum system

In order to answer this question whether a nucleus is a classical or quantum system, let us investigate if the wave nature of matter is relevant in a nucleus. As a general rule, the wave nature of matter is relevant where the de Broglie wavelength associated with the particles is of the order of the size of the system. The de Broglie wavelength of a nucleon of energy 10 MeV is

$$\lambda = \frac{h}{\sqrt{2mE}} = \frac{6.626 \times 10^{-34}}{[2 \times 1.67 \times 10^{-27}(10 \times 10^6)(1.602 \times 10^{-19})]^{1/2}}$$

$$= 9.06 \times 10^{-15} \text{ m} = 9.06 \text{ fm}$$

Since it is of the order of the size of a nucleus, the wave nature of matter is applicable in case of a nucleus. Thus, the motions of nucleons in a nucleus are governed by the laws of quantum physics. However, the classical pictures in which nucleons are treated as little balls moving around - applied so successfully in describing gasses or liquids - are of limited usefulness.

6. Packing fraction

Packing fraction, denoted by f, of a nucleus is expressed as

$$f = \frac{\text{Atomic mass} - \text{Mass number}}{\text{Mass number}} = \frac{m(Z, A) - A}{A}$$

where $m(Z, A)$ is the actual weight of the nucleus in amu, and A is the mass number. The mass number A is an integer. Thus, we have

$$m(Z, A) = A(1 + f)$$

The variation of packing fraction f versus mass number A is shown in Figure 2. As A increases, the packing fraction becomes negative, passes through a rather flat minimum and then rises gradually, becoming positive again at about 180 mass number.

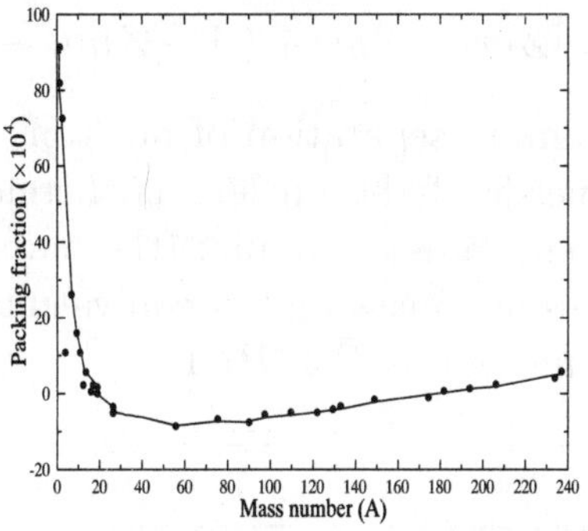

Figure 2: Packing fraction as a function of mass number.

7. Binding energy

One may at first assume that the mass of an atom should be equal to the sum of the masses of its constituent particles, *i.e.*, electrons, protons and neutrons. But, the mass of a stable atom was found to be less than the sum of the masses of its constituent particles. The decrease in the mass is known as the 'mass defect'. This mass defect can be explained in the following manner. The nucleons in a stable nucleus are held together by strong attractive forces, and therefore some work need to be done in separating them from one another until they are at very large distances from one another. In other words, the energy needs to be supplied to a nucleus in order to break it into its constituents, separated by very large distance from one another. The principle of equivalence of mass and energy confirms the idea that the amount of energy required to break up the atom into its constituents is equivalent to the mass defect. It can also be defined as the amount of mass which would be converted into the energy if the particular atom is to be assembled out of its constituents. The energy equivalent of mass defect is thus a measure of the binding energy of the atom.

The constituents of the atom $_Z X^A$ are Z protons, Z electrons and $(A - Z)$ neutrons. The total mass of individual constituents is $Zm_e + Zm_p + (A-Z)m_n$, where m_e, m_p and m_n are, respectively, the masses of electron, proton and neutron. If $m(Z, A)$ is the mass of the atom $_Z\mathrm{X}^A$, then the mass defect Δm is

$$\Delta m = [Zm_p + Zm_e + (A - Z)m_n] - m(Z, A)$$

As the binding energy B is the energy equivalent of the mass defect, we have

$$B = c^2 \Delta m = [Zm_p + Zm_e + (A - Z)m_n - m(Z, A)]c^2$$

Since the state of infinite separation of nucleons is taken to be at zero energy, the total energy of the nucleus is therefore $-B$. The binding energy may either be expressed in amu (the mass unit) or in MeV (the energy unit). One atomic mass unit is equivalent to 931.478 MeV. The binding energy per nucleon is $\overline{B} = B/A$.

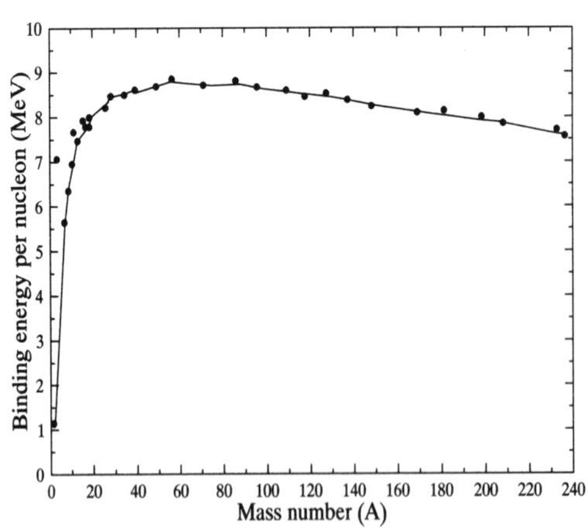

Figure 3: Binding energy per nucleon versus mass number.

A graph of binding energy per nucleon $\overline{B} = B/A$ versus mass number A is shown in Figure 3. The graph gives us following valuable information:

(a) With the exception of He^4, Be^8, C^{12} and O^{16}, the values fall on or in the proximity to a single curve. The binding energies of some of the lighter nuclides, such as H^1, H^2 and He^3 are very low. However, over a very considerable range of mass numbers, the binding energy per nucleon is close to 8 MeV. This explains the energy release during the fusion of light elements.

(b) The positive value of B/A for all values of A shows the stability of nuclei. This shows that there are attractive nuclear forces among the constituent particles within the nuclear volume.

(c) For very light nuclei, the value of B/A versus A varies in an erratic manner. There are notable peaks at $A = 4n$ (He^4, Be^8, C^{12} O^{16}, Ne^{20}, Mg^{24}). This reflects the peculiar stability of the α-particle structure. This great stability is mainly a consequence of the pairing of the two protons and two neutrons with opposite spins.

(d) Above $A \sim 20$, the binding energy per nucleon B/A is remarkably constant (within 10% variation). This must reflect an important property of nuclear forces. If each of the nucleons A inside the nucleus interacts with all the remaining $(A-1)$ nucleons with the same strength, one would expect B to increase as $A(A-1)$ or approximately, as A^2 for heavy nuclei. But, agreement with the observations is obtained when it is postulated that the nucleons interact with only a limited number of neighbours around the nucleon or the nuclear forces are saturated.

(e) The curve indicates that B/A increases to a maximum value of about 8.8 MeV per nucleon around the mass number 56 and then decreases slowly to about 7.6 MeV per nucleon at $A = 238$. This explains the reason for the energy release during the fission process of a heavy nucleus.

(f) Discontinuities in the curve of B/A versus A can be noticed at the points where at least one of the Z or $(A - Z)$ is equal to the magic number, 2, 8, 20, 28, 50, 82, 126 or 184.

7.1 Semi-empirical mass formula

The mass $m(Z, A)$ of a neutral atom having Z protons, Z electrons and $(A - Z)$ neutrons is

$$m(Z, A) = Zm_p + Zm_e + (A - Z)m_n - B/c^2$$

where the binding energy B may be regarded as made up of a number of terms, each of which representing some general characteristic of the nucleus. von Weizsäcker obtained a formula, known as the semi-empirical mass formula or Bethe-Weizsäcker mass formula, for expressing the atomic mass of a nuclide, in terms of a series of binding energy terms. It is of great importance in nuclear stability problems. We shall now discuss about various terms contributing there.

(i) Volume energy

Since the nuclear density is constant and nuclear volume is proportional to the number of nucleon in it, it is reasonable to assume a volume energy E_v that binds the nucleons in the nucleus is due to inter-nucleon interaction:

$$E_v \propto \frac{4}{3}\pi R^3 \propto \frac{4}{3}\pi R_0^3 A = a_v A$$

Here, we have used the relation $R = R_0\ A^{1/3}$. The coefficient a_v is the energy coefficient representing the binding energy per nucleon.

(ii) Surface binding energy

The nucleons which lie well within the nucleus experience the full attraction by the surrounding nucleons but the nucleons that lie on the surface or near the surface do not experience the full attractive force and so the volume energy overestimates the binding energy. If R is the nuclear radius and T the surface tension, then the surface energy E_s is expressed as

$$E_s = 4\pi R^2 T = 4\pi R_0^2 A^{2/3} T = a_s A^{2/3},$$

where the surface energy coefficient $a_s = 4\pi R_0^2 T$.

(iii) Coulomb energy

Coulomb energy result since a charge of Ze is present within the volume of a nucleus. For a homogeneously charged liquid drop with radius R and charge density

$$\rho_c = \frac{Ze}{4\pi R^3/3}$$

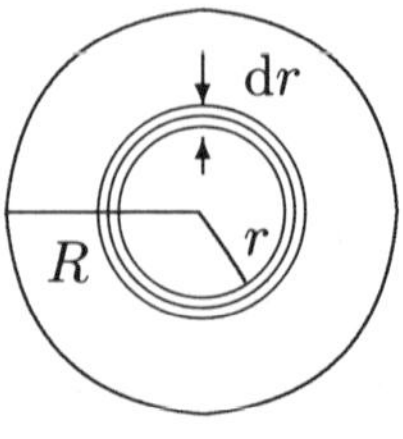

we can evaluate the Coulomb contribution of the nuclear binding energy with the help of classical argument. Suppose, we add a spherical shell of thickness dr from outside to the sphere of radius r, then the Coulomb energy needed is expressed as

$$E_c = \int_0^R \frac{(\frac{4}{3}\pi r^3 \rho_c)(4\pi r^2 \mathrm{d}r \rho_c)}{r}$$

For the above charge density, the integral gives

$$E_c = \frac{3}{5}\frac{Z^2e^2}{R} = \frac{3}{5}\frac{Z^2e^2}{R_0 A^{1/3}} = a_c\frac{Z^2}{A^{1/3}}$$

where the coulomb energy coefficient $a_c = (3e^2/5R_0)$. As the Coulomb force is repulsive, its contribution to the binding energy is negative.

Note: The volume energy, surface energy and Coulomb energy which are considered above are purely classical concepts that arise from the liquid drop model of nucleus. In addition to them, we should the quantum mechanical concepts of quantized energy levels and the application of Pauli's exclusion principle for filling the energy states and also the concept of pairing of spins.

(iv) Asymmetry energy

For stability of a nucleus of given mass number A, the energy should be the minimum and this can be achieved by putting protons and neutrons as many as possible in the lower possible energy levels in accordance with the Pauli's exclusion principle. (Remember that protons and neutrons both are fermions.) For a stable nucleus, the number of protons should be equal the number of neutrons. If there is an excess of protons or neutrons, then the nucleus will be unstable. In other words, the excess number of protons $(Z - N)$ or of neutrons $(N - Z)$ will have less binding energy and consequently, we have to introduce an additional term, $a_a(N - Z)^2/A$, known as the asymmetry energy. Here, a_a is the asymmetry coefficient.

(v) Pairing energy

It is observed that a nucleus with even number of protons and even number of neutrons (even-even nucleus) is more stable than even-odd or odd-even nucleus. An odd-odd nucleus is more unstable than the even-odd or odd-even nucleus. This is due to the pairing of spins. A nucleus with paired spins of protons and neutrons will be more stable than the nucleus with unpaired spins. This effect has been incorporated in the binding energy formula by considering the pairing energy term δ In the following form.

$$\delta = \begin{cases} +a_p/A^{3/4} & \text{even-even nucleus} \\ 0 & \text{odd-even or even-odd nucleus} \\ -a_p/A^{3/4} & \text{odd-odd nucleus.} \end{cases}$$

After inclusion of these two additional terms, the binding energy is expressed as the following.

$$B = a_v A - a_s A^{2/3} - a_c \frac{Z^2}{A^{1/3}} - a_a \frac{(N-Z)^2}{A} + \delta \tag{1.2}$$

It is possible to estimate the values of the coefficients in equation (1.2) from theoretical considerations but it is found more convenient to treat them as empirical coefficients and determine them with the help of the data of masses of stable Nuclides and energetics of nuclear reactions. Several sets of values for these coefficients are available in literature and one of them is as the following.

$$a_v = 15.75 \text{ MeV}, \qquad a_s = 17.8 \text{ MeV}, \qquad a_c = 0.71 \text{ MeV},$$

$$a_a = 23.70 \text{ MeV}, \qquad a_p = 33.5 \text{ MeV} \tag{1.3}$$

7.2 Application of empirical mass formula

(i) Most stable isobar

Bethe-Weizsäcker mass formula is

$$\begin{aligned} m(Z,A) &= Zm_p + Zm_e + (A-Z)m_n - a_v A + a_s A^{2/3} + a_c \frac{Z^2}{A^{1/3}} \\ &\quad + a_a \frac{(N-Z)^2}{A} - \delta \\ &= Zm_p + Zm_e + (A-Z)m_n - a_v A + a_s A^{2/3} \\ &\quad + a_c \frac{Z^2}{A^{1/3}} + a_a \frac{(A-2Z)^2}{A} - \delta \end{aligned} \tag{1.4}$$

Equation (1.4) is obtained by replacing N by $A - Z$. Let us find out the most stable isobar. For this, we differentiate expression (1.4) (omitting the pairing term δ) with respect to Z and put it equal to zero.

$$\left(\frac{\partial m}{\partial Z}\right)_{A=\text{const}} = m_p + m_e - m_n + a_c \frac{2Z}{A^{1/3}} - 4a_a \frac{A-2Z}{A} = 0 \tag{1.5}$$

If $Z = Z_0$ satisfies the relation (1.5), then we have

$$2a_c\frac{Z_0}{A^{1/3}} - 4a_a\frac{A-2Z_0}{A} = m_n - m_p - m_e$$

and

$$Z_0 = \left\{\frac{(m_n - m_p - m_e) + 4a_a}{2a_cA^{2/3} + 8a_a}\right\}A \tag{1.6}$$

For the given data, $m_n = 939.79$ MeV, $m_p = 938.59$ MeV, $m_e = 0.51$ MeV, we get $Z_0 = 43.8$ for the stable isobar with $A = 101$, using values (1.3) for the coefficients. Since the mass number is an integer, we have to choose the nearest integer obtained for Z_0 using equation (1.6). The nucleus $Z_0 = 44$ is the most stable isobar for $A = 101$. Since this is an odd nucleus, the pairing term $\delta = 0$ and we obtain only one curve.

For $A = 106$, equation (1.6) gives the value $Z_0 = 45.78$. The nearest integer is 46. This is an even-even nucleus for which one has to include the pairing term. An even A nucleus can also be an odd-Z odd-N nucleus, for which also the pairing energy is to be included. Thus, one curve is for even-even nucleus and the other for odd-odd nucleus.

The said curves are parabolas. This is because the expression (1.4) is quadratic in Z. Omitting the pairing term δ, equation (1.4) can be written as

$$m(A,Z) = \alpha A + \beta Z + \gamma Z^2, \tag{1.7}$$

with

$$\alpha = m_n - a_v + a_a + \frac{a_s}{A^{1/3}}$$

$$\beta = m_p + m_e - m_n - 4a_a$$

$$\gamma = a_cA^{-1/3} + \frac{4}{A}a_a.$$

Equation (1.7) is a quadratic equation in Z and hence yields a parabola for the curve connecting $m(A,Z)$ and Z for a constant A.

(ii) Nuclear fission

When a heavy nucleus is bombarded by a neutron, it is found that it splits into two roughly equal nuclei with the emission of two or more neutrons. This phenomenon is known as the 'nuclear fission'. Let us consider the spontaneous fission of $^{236}U_{92}$ and calculate the energy released using the Bethe-Weizsäcker mass formula (1.4). The energy released may roughly estimated as

$$Q = m(A,Z) - 2m(A/2, Z/2)$$

The Q-value can easily be calculated using equation (1.4). The volume term and the asymmetry term do not contribute to the Q-value, as we are considering binary fission. Neglecting the pairing term δ, we get

$$Q = -2a_s\left(\frac{A}{2}\right)^{2/3} - 2a_c\left(\frac{Z}{2}\right)^2\left(\frac{A}{2}\right)^{-1/3} + a_sA^{2/3} + a_cZ^2A^{-1/3} \quad (1.8)$$

$$= a_sA^{2/3}\left\{1-(2)^{1/3}\right\} + a_cZ^2A^{-1/3}\left\{1-(2)^{-2/3}\right\} \quad (1.9)$$

$$= -0.26a_sA^{2/3} + 0.37a_cZ^2A^{-1/3}. \quad (1.10)$$

From equation (1.10), the Q-value for ${}_{92}U^{236}$ can be calculated easily after using $a_s = 17.8$ MeV and $a_c = 0.71$ MeV. It comes out to be $Q = 183$ MeV. This is the amount of energy released per fission of ${}_{92}U^{236}$. When the Q-value exceeds the Coulomb potential barrier, then the fission takes place spontaneously. The Coulomb potential barrier is the repulsive potential between the two fissioning fragments at the point of separation.

$$\text{Coulomb potential barrier} = \frac{(Z/2)^2e^2}{2r} = \frac{(Z/2)^2e^2}{2r_0(A/2)^{1/3}} = \frac{Z^2e^2}{r_0}\frac{2^{1/3}}{8A^{1/3}}$$

The Coulomb energy of the parent nucleus is

$$E_c = \frac{3}{5}\frac{Z^2e^2}{R} = \frac{3}{5}\frac{Z^2e^2}{r_0A^{1/3}} = a_c\frac{Z^2}{A^{1/3}}$$

with $a_c = 3e^2/5r_0$. To find the condition for spontaneous fission, it is advantageous to express the Coulomb potential barrier in terms of a_c.

$$\text{Coulomb potential barrier} = \frac{5}{24}(2)^{1/3}\,a_c\,\frac{(Z)^2}{(A)^{1/3}} = 0.262a_c\frac{Z^2}{A^{1/3}} \quad (1.11)$$

From equations (1.10) and (1.11), we obtain the condition for spontaneous fission.

$$-0.26a_sA^{2/3} + 0.37a_cZ^2A^{-1/3} > 0.262a_c\frac{Z^2}{A^{1/3}} \quad \text{or} \quad \frac{Z^2}{A} > 2.4\,\frac{a_s}{a_c}$$

Substituting the values as $a_s = 17.8$ MeV and $a_c = 7.1$ MeV, we find the condition for spontaneous symmetric fission as

$$\frac{Z^2}{A} > 60$$

The quantity Z^2/A is known as the 'fissionability parameter'. The above value of fissionability parameter is only a rough estimate since it does not take into account the tunneling effect which is a finite probability of penetrating the potential barrier. The above discussion only points out that the fissionability depends upon the interplay between the forces associated with the surface energy and the Coulomb energy, and that the fission is favoured only in heavy nuclei and not in the light nuclei.

Exercise 3: Calculate the binding energy of a deuteron ${}_1H^2$. The masses of a deuteron, a proton and a neutron in amu are 2.014102, 1.007825 and 1.008665, respectively. Given 1 amu = 931 MeV.

Solution: Mass of one proton and one neutron = 1.007825 + 1.008665 = 2.01649 amu

Mass defect $\Delta m = 2.01649 - 2.014102 = 0.002388$ amu

Binding energy of deuteron = $0.002388 \times 931 = 2.223$ MeV

Exercise 4: Calculate the binding energy of the following isobars and their binding energies per nucleons.

${}_{28}Ni^{64} = 63.9280$ amu ${}_{29}Cu^{64} = 63.9298$ amu

Which of these would you expect to be β^- active and how would it decay? Why?

Solution: (a) Since mass of neutron $m_n = 1.008665$ amu and $m_p = 1.007825$ amu, hence for ${}_{28}Ni^{64}$, we have

$$Nn_n + Zm_p = 36 \times 1.008665 + 28 \times 1.007825 = 64.531 \text{ amu}$$

Thus,

$$\Delta m = 64.531 - 63.9280 = 0.603 \text{ amu}$$

Hence, the binding energy = 561.4 MeV and binding energy per nucleon is 8.77 MeV.

For ${}_{29}Ni^{64}$, we have

$$Nm_n + Zm_p - m(Z, A) = 35 \times 1.008665 + 29 \times 1.007825 = 64.5272 \text{ amu}$$

Thus,

$$\Delta m = 64.5272 - 63.9298 = 0.5974 \text{ amu}$$

Hence, the binding energy = 556.18 MeV and binding energy per nucleon is 8.7 MeV.

(b) These isobars are β^- emitters, *i.e.*,

${}_{28}Ni^{64} \rightarrow {}_{29}Cu^{64} + \beta^-$ and ${}_{29}Cu^{64} \rightarrow {}_{30}Zn^{64} + \beta^-$

when the mass of of the final isotope is less than the corresponding initial isotope *i.e.*, $m(Z, A) > m(Z+1, A)$. Since the mass of ${}_{29}Cu^{64}$ is greater than that of ${}_{28}Ni^{64}$ and ${}_{30}Zn^{64}$, hence ${}_{28}Ni^{64}$ is not a β^- emitter but ${}_{29}Cu^{64}$ is a β^- emitter.

These isobars are β^+ emitter, if $m(Z, A) > m(Z-1, A) + 2\ {}_1e^0$

Since m (Cu^{64}) - m (Ni^{64}) $= 0.0018$ amu $= 1.6758$ MeV is greater than the mass of two electrons, hence Cu^{64} is a β^+ emitter.

Exercise 5: Calculate the amount of energy released when two ${}_1H^2$ nuclei fuse together to form a ${}_2He^4$ nucleus. The binding energy per nucleon of H^2 and He^4 is 1.1 MeV and 7.0 MeV, respectively.

Solution: As the number of nucleon in ${}_2He^4$ is 4, the binding energy for ${}_2He^4$ is 28.0 MeV. Similarly the binding energy of ${}_1H^2$ is 2.2 MeV.

Mass of ${}_2He^4$ nucleus = 2 [Mass of proton] + 2 [Mass of neutron] - 28.0 MeV

Mass of ${}_1H^2$ nucleus = [Mass of proton] + [Mass of neutron] - 2.2 MeV

The amount of energy released during the fusion reaction is

$\Delta E = 2$ [Mass of ${}_1H^2$] - [Mass of ${}_2He^4$]

$= 2\ [m_p + m_n - 2.2] - [2\ m_p + 2\ m_n - 28.0] = 23.6$ MeV

8. Problems and questions

1. Discuss about the objections put forth for the existence of electron in a nucleus.

2. Discuss about the proton-neutron theory of a nucleus.

3. Discuss about the size, density, charge distribution in a nucleus.

4. Show that a nucleus is a quantum system.

5. Describe about the binding energy of a nucleus.

6. Discuss about the liquid drop model of a nucleus

7. Write notes on the following

   (i) Proton-electron theory of a nucleus

   (ii) Proton-neutron theory of a nucleus

   (iii) Energies of nucleons in a nucleus

(iv) Isotopes, isotones, isobars, isodiapheres, isomers, mirror nuclei

(v) Packing fraction

(vi) Binding energy

(vii) Difference between spin and iso-spin

II. Nuclear Forces

A nucleus in general consists of a number of protons and neutrons. A proton carries positive charge and therefore there is electrostatic repulsion between two protons. A neutron carry no electric charge and thus there is no coulomb force between two neutrons as well as between a proton and a neutron. In spite of the coulomb repulsion among the protons, the protons along with neutrons are confined in a nucleus. It leads to an idea that there should be some kind of force of the attraction between the nucleons. In order to understand about the force of attraction between the nucleons, let us look into a graph (Figure 1) showing variation of the number of neutrons versus the number of protons in the stable nuclei.

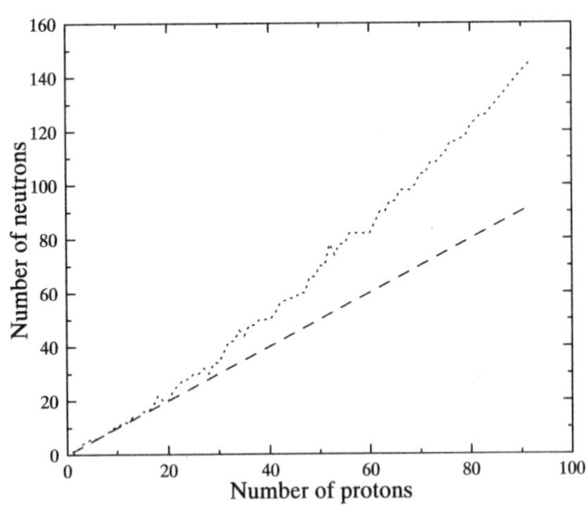

Figure 1: Variation of the number of neutrons versus the number of protons in the stable nuclei. Dotted line is drawn for equal number of neutrons and protons.

We know that hydrogen is the only nucleus which does not have any neutron. It has only one proton. For the lighter nuclei, the number of neutrons is approximately equal to the number of protons. With the increase of the mass of a nucleus, the ratio of the number of neutrons to the number of protons increases and becomes very large for heavy nuclei. The equality of the number of protons to the number of neutrons for the light nuclei can be interpreted as showing that there is a strong attractive force between a neutron and a proton. The stability of deuteron, which is made up of one neutron and one proton, supports this conclusion. It shows that this kind of force does not depend on the electric charge of nucleons. When there is a strong nuclear attractive force between a neutron and a proton, it appears reasonable to assume that there are similar forces: (i) between two neutrons and (ii) between two protons. Let us denote the forces between a neutron and a proton, between two neutrons and between two protons as n - p, n - n and p - p, respectively. Some conclusions which may be derived from Figure 1. are as the following:

(i) In case the force between two neutrons (n - n) had been greater than that n - p or p - p, the most abundant, stable light nuclei would have more neutrons than the protons. But, in the most abundant light nuclei, the number of neutrons is equal to the number of protons. Hence, n - n is not larger than n - p as well as p - p.

(ii) In case the force between two protons (p - p) had been greater than that n - p or n - n, the most abundant, stable light nuclei would have more protons than the neutrons. But, in the most abundant light nuclei, the number of neutrons is equal to the number of protons. Hence, p - p is not larger than n - p as well as n - n.

(iii) The increase of the number of neutrons with the increase of the number of protons can be interpreted that there is a force between two neutrons.

(iv) With the increase of the number of protons in a nucleus, the coulomb repulsion increases. On the other side, the ratio of the number of neutrons to that of protons increases gradually from 1 (for the light nuclei) to 1.6 for uranium ($Z = 92$). This increase in the ratio may be interpreted as the necessary requirement for increasing the net attractive force, which is required to balance the electrostatic repulsion between the protons.

Since the coulomb force is a long range force, the force of attraction between the nucleons should be of short range in nature.

In view of the above discussion, it was concluded that there are forces of attraction between each pair of nucleons, irrespective of electric charge on them. The n-n and p-p forces are assumed to be nearly equal in magnitude and to be not greater than that of the n-p force. They may however be much smaller than the n-p force, or they may have the same, or very nearly the same magnitude, as the n-p force. The two possibilities may be expressed as:

I. $\quad n\text{-}p \approx n\text{-}n \approx p\text{-}p$

II. $\quad n\text{-}p >> n\text{-}n; \qquad n\text{-}p >> p\text{-}p; \qquad n\text{-}n \approx p\text{-}p$

Wagner in a general treatment of nuclear energy levels, made the assumption that the relationship described by case I holds, that is the nuclear forces n-p, n-n and p-p are equal in magnitude. This assumption is known as the hypothesis of 'charge independence' of nuclear forces. The alternative assumption that the n-n and p-p forces are equal to the n-p force is known as the hypothesis of 'charge symmetry' of nuclear forces. It is less restrictive than the hypothesis of charge independence.

1. Yukawa potential

In order to understand about the forces existing in a nucleus, Yukawa postulated a new type of field, called the 'meson field'. For the electromagnetic field between two charged particles, it was found quantum mechanically that the force between the particles was due to the exchange of photons between them. Yukawa extended this analogy and assumed the exchange of mesons between the nucleons. Thus, Yukawa assumed that the force between two nucleons was due to the exchange of mesons between them. It was assumed that a nucleon (proton as well as neutron) consists of some kind of common core surrounded by rapidly pulsating meson cloud. Since mesons (called pions) may have positive, negative and zero charge, the rapid back and forth jumping of mesons between a pair of nucleons binds the two nucleons together. Let us consider the case of the nucleus of deuteron, where the spins of both proton and neutron are parallel to each other so that the total spin is 1. The constant charge exchange between the nucleons gives a strong force of attraction between them.

We know that the electrostatic potential between two protons, each having electric charge e and separated by a distance r is

$$V = -\frac{e^2}{r} \tag{2.1}$$

This is a long range force and becomes zero at infinite distance. Such form cannot be valid for the nuclear force, as the nuclear force between two nucleons is a short range force. In the analogy with the electric charge e, suppose a nucleon carries a mesonic charge f. (This mesonic charge is quite different in nature from the electric charge e on the proton. Further, a neutron also carries the mesonic charge f.) Both the proton and neutron have this mesonic charge f. For the nuclear force, Yukawa proposed a potential $V(r)$ of the form

$$V(r) = -V_0 \frac{\mathrm{e}^{-r/a}}{r} \qquad \text{Yukawa well type} \tag{2.2}$$

where r is the separation between two nucleons, a having the dimensions of length is equal to 2×10^{-15} m. Here, the exponential factor creates the fact that the nucleons exert forces on their immediate neighbours only. It shows a short range nature of the nuclear force. Other expressions showing short range nature of potential could be, for example,

$$V(r) = -V_0 \, \mathrm{e}^{-(r/a)^2} \qquad \text{Gaussian well type}$$

and

$$V(r) = -V_0 \, \mathrm{e}^{-(r/a)} \qquad \text{Exponential well type}$$

But the expression (2.2) could reproduce the results more successfully and therefore it was accepted by the scientific community. Extending the analogy of electromagnetic waves produced by an accelerated electric charge, Yukawa assumed that the mesonic charge, oscillating between two nucleons produces a mesonic wave with frequency ν_m expressed as

$$\nu_m = c\Big(\overline{\nu}_m^2 + \frac{1}{4\pi^2 a^2}\Big)^{1/2} \tag{2.3}$$

where $\overline{\nu}_m$ is the wave number of the mesonic wave. In the case of a photon, the second term would be absent (*i.e.*, $a = \infty$).

In analogy with a photon, the rest mass m_0 of the meson can be derived by comparing the expressions for the total energy E from the wave and particle aspects. From the particle point of view, the relativistic energy E is expressed as

$$E^2 = m_0^2 c^4 + p^2 c^2 \tag{2.4}$$

From the wave point of view,

$$E = h\nu_m \qquad \text{and} \qquad p = h/\lambda = h\overline{\nu}_m \tag{2.5}$$

Substituting ν_m and $\overline{\nu}_m$ from equation (2.5) in equation (2.3), we get

$$E^2 = c^2\Big[p^2 + \Big(\frac{h}{2\pi a}\Big)^2\Big] \tag{2.6}$$

Comparison of equations (2.4) and (2.6) gives

$$m_0^2c^4 = c^2\left(\frac{h}{2\pi a}\right)^2 \qquad \text{or} \qquad m_0 = \frac{h}{2\pi ca} \tag{2.7}$$

Remember that the rest of a photon is zero. Substituting the values of h, a and c in equation(2.7), the rest mass of meson is

$$m_0 = \frac{6.626 \times 10^{-34}}{2 \times 3.14 \times 3 \times 10^8 \times 2 \times 10^{-15}} = 1.76 \times 10^{-28} \quad \text{gm}$$

Thus, the rest mass of meson is about 200 times more than the rest mass of an electron.

A particle which appeared to satisfy the predicted properties of Yukawa meson was discovered in 1936, and is known as the μ-meson with rest mass 207 times electron mass. However, from the detailed analysis it was found that this particle could not produce the nuclear forces of the right magnitude. This discrepancy was later on overcome by the discovery of π-meson in 1947, having a mass 270 times the electron mass. Its characteristics of this particle were found to be in agreement with the predicted particle.

Based on the Yukawa potential between two nucleons, the attraction between a nuclear proton and a neutron can be represented as shown in the following picture.

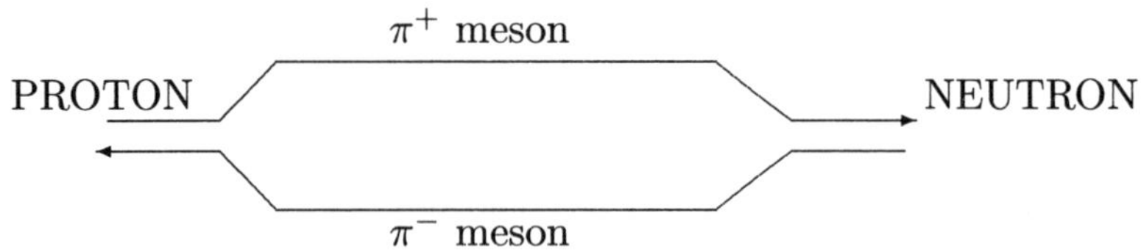

Figure 2: Proton-neutron exchange.

The particles in the nucleus are held together by the continuous exchange of mesons, expressed as the following.

$$n \rightarrow p + \pi^- \qquad \text{and} \qquad p \rightarrow n + \pi^+$$

1.1 Exchange Force

In order to explain the saturation of nuclear forces in 1932, Heisenberg proposed that nuclear forces were the exchange forces, which would depend explicitly on the symmetry of the wave function. At the time of Heisenberg's idea of exchange forces, mesons were known and it was known that π-mesons were being exchanged between nucleons, by any of the following process:

1. $(n, p) \rightarrow (p + \pi^-, p) \rightarrow (p, \pi^- + p) \rightarrow (p, n)$

2. $(p, n) \rightarrow (n + \pi^+, n) \rightarrow (n, \pi^+ + n) \rightarrow (n, p)$

3. $(p, p) \rightarrow (p + \pi^0, p) \rightarrow (p, p + \pi^0) \rightarrow (p, p)$

4. $(n, n) \rightarrow (n + \pi^0, n) \rightarrow (n, n + \pi^0) \rightarrow (n, n)$

The exchange of a pion is thus equivalent to the exchange of charge.

2. Deuteron

The simplest example of a nucleus where both proton and neutron are available is the deuteron. The deuteron has one proton and one neutron. Since the deuteron has measurable values of nuclear parameters, it may be used as a good example for understanding the nuclear forces. Some properties of a deuteron are as the following:

1. A deuteron consists of one proton and one neutron, which have nearly equal masses, $m_p \approx m_n = m$, so that the reduced mass of deuteron is $m/2$. Here, m_p stands for the mass of proton and m_n for the mass of neutron.

2. The binding energy of deuteron is about 2.225 MeV, which is much smaller than 8 MeV, as needed to pull a nucleon out of a medium-mass nucleus. Therefore, a deuteron may be treated as a loosely bound system.

3. Experimental results suggest that the spins of proton and neutron in the ground state of deuteron are parallel to each other ($I = 1$). Thus, the ground state is a triplet state. The orbital angular momentum of deuteron about the centre of mass is zero. Thus, the ground state of deuteron is ^{3}S.

4. The parity of the ground state of deuteron, as measured indirectly, is found to be even (positive).

5. The magnetic dipole moments of proton and neutron, obtained from measurements, are

$$\mu_p = 2.7925 \frac{eh}{4\pi m_p} \qquad \text{and} \qquad \mu_n = -1.9128 \frac{eh}{4\pi m_p}$$

When the deuteron is assumed to be in the ground state with $l = 0$, there would be no contribution from the orbital motion. Thus, one would expect the magnetic moment of deuteron to be equal to the sum of the magnetic moments of proton and neutron:

$$\mu_p + \mu_n = 0.8797 \; \frac{eh}{4\pi m_p}$$

However, from the measurements, the magnetic moment of deuteron is found as

$$\mu_d = 0.8574 \; \frac{eh}{4\pi m_p}$$

Though the difference between the measured and expected values is not very large and there are some uncertainties, but the simplest explanation for the difference may be that there could be some orbital motion and therefore the assumption of the ground state with $l = 0$ is not completely correct.

There is a even better evidence for this conclusion, derived from the measurement of the electric quadrupole moment of deuteron, which is found to be

$$Q_d = 2.82 \times 10^{-27} \text{ cm}^2$$

Though this value is relatively small in comparison to others, but is non-zero.

3. Schrödinger equation for deuteron

From the point of view of quantum mechanics, deuteron is considered as a stationary state formed by mutual attraction between a proton and a neutron. We are aware that a two-body system can be reduced to a one-body system where a fictitious body of reduced-mass μ moves relative to the centre of mass of the system. The Schrödinger wave equation for deuteron in the centre of mass system can be expressed as

$$\nabla^2\psi + \frac{2\mu}{\hbar^2}\Big(E - V\Big)\psi = 0 \tag{2.8}$$

where the reduced mass μ of deuteron is defined as

$$\frac{1}{\mu} = \frac{1}{m_p} + \frac{1}{m_n} \qquad \text{or} \qquad \mu = \frac{m_p m_n}{m_p + m_n}$$

where m_p and m_n are masses of proton and neutron, respectively. As the difference between the masses of proton and neutron is very small, we can take $m_p \approx m_n = m$, so that $\mu = m/2$. E is the total energy of the system and V the potential energy describing the force between the two bodies. In terms of spherical polar coordinates, equation (2.8) becomes

$$\left[\frac{1}{r^2}\frac{\partial}{\partial r}\left(r^2\,\frac{\partial\psi}{\partial r}\right)+\frac{1}{r^2\sin\theta}\frac{\partial}{\partial\theta}\left(\sin\theta\,\frac{\partial\psi}{\partial\theta}\right)+\frac{1}{r^2\sin^2\theta}\frac{\partial^2\psi}{\partial^2\phi}\right]$$

$$+\frac{2\mu}{\hbar^2}\left[E-V(r)\right]\psi=0 \qquad (2.9)$$

For convenience, here we consider the potential energy V to be spherically symmetric, so that it depends only on the radial distance r. This simplification helps in breaking the Schrödinger equation into two equations, one depending on r only (called the radial part) and the other depending on θ and ϕ only (called the angular part). The solution (wave function) of the above equation can be written as the product of two functions, one depending on r only and the other on θ and ϕ only, as $\psi(r,\theta,\phi) = R(r)\ Y(\theta,\phi)$. Substituting this expression in equation (2.9) and rearranging we get

$$\frac{1}{R}\frac{\mathrm{d}}{\mathrm{d}r}\left(r^2\,\frac{\mathrm{d}R}{\mathrm{d}r}\right)+\frac{2\mu r^2}{\hbar^2}\left[E-V\right]=-\frac{1}{Y}\left[\frac{1}{\sin\theta}\frac{\partial}{\partial\theta}\left(\sin\theta\,\frac{\partial Y}{\partial\theta}\right)+\frac{1}{\sin^2\theta}\frac{\partial^2 Y}{\partial\phi^2}\right]$$

The left hand side of this equation depends only on r, whereas the right hand side depends only on θ and ϕ. For all values of the variables, each side of this equation must separately be equal to the some constant, which comes out to be $l(l+1)$, where l is the angular momentum quantum number for the system. Thus we have

$$\frac{1}{r^2}\frac{\mathrm{d}}{\mathrm{d}r}\left(r^2\,\frac{\mathrm{d}R}{\mathrm{d}r}\right)+\frac{2\mu}{\hbar^2}\left[E-V(r)-\frac{l(l+1)\hbar^2}{2\mu r^2}\right]R=0 \qquad (2.10)$$

and

$$\frac{1}{\sin\theta}\frac{\partial}{\partial\theta}\left(\sin\theta\,\frac{\partial Y}{\partial\theta}\right)+\frac{1}{\sin^2\theta}\frac{\partial^2 Y}{\partial\phi^2}+l(l+1)Y=0 \qquad (2.11)$$

The equation (2.11) is well known and its solution is a spherical harmonics Y_{lm}, where l is the orbital quantum number and m the magnetic quantum number.

3.1 Ground state of deuteron

For the ground state ^{3}S with ($l = 0$), we have $Y_{00} = 1/\sqrt{4\pi}$ and equation (2.10) gives

$$\frac{1}{r^2}\frac{\mathrm{d}}{\mathrm{d}r}\left(r^2\frac{\mathrm{d}R}{\mathrm{d}r}\right) + \frac{2\mu}{\hbar^2}\Big[E - V(r)\Big]R = 0 \tag{2.12}$$

For the deuteron, the reduced mass $\mu = m/2$ and substituting $R = u(r)/r$ in equation (2.12), where $u(r)$ is called the radial wave function, we have[1]

$$\frac{\mathrm{d}^2u}{\mathrm{d}r^2} + \frac{m}{\hbar^2}\Big[E - V(r)\Big]u = 0 \tag{2.13}$$

Here, the potential $V(r)$ describes the force of attraction between a pair of nucleons. It is a short range force which decreases with the increase of the distance r and vanishes for practical purposes when $r > b$, where b is known as the range of the nuclear force.

The actual form of the nuclear potential is unknown. For convenience, scientists generally assume it to be of square-well type with the depth V_0 and width b as shown in Figure 3. It can mathematically expressed as

$$V(r) = \begin{cases} -V_0 & \text{for} \quad 0 \le r < b \\ 0 & \text{for} \quad r > b \end{cases}$$

There seems to be no reason to believe that the potential is a square-well type. The only point is that this expression is very simple and for it,

---

[1]We have $R = u/r$, so that

$$\frac{\mathrm{d}R}{\mathrm{d}r} = \frac{1}{r}\frac{\mathrm{d}u}{\mathrm{d}r} - \frac{u}{r^2} \qquad \text{and} \qquad r^2\frac{\mathrm{d}R}{\mathrm{d}r} = r\frac{\mathrm{d}u}{\mathrm{d}r} - u$$

Thus,

$$\frac{\mathrm{d}}{\mathrm{d}r}\left(r^2\frac{\mathrm{d}R}{\mathrm{d}r}\right) = r\frac{\mathrm{d}^2u}{\mathrm{d}r^2} \qquad \text{and} \qquad \frac{1}{r^2}\frac{\mathrm{d}}{\mathrm{d}r}\left(r^2\frac{\mathrm{d}R}{\mathrm{d}r}\right) = \frac{1}{r}\frac{\mathrm{d}^2u}{\mathrm{d}r^2}$$

the Schrödinger equation can be solved exactly.

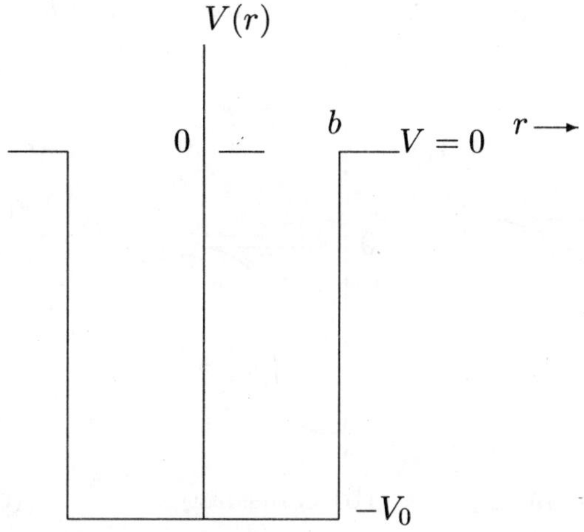

Figure 3: Square-well potential for deuteron.

It is not out of context to state here that few other expressions for the potential, given in Figure 4 and expressed as:

$$V(r) = -V_0\ \mathrm{e}^{-(r/a)^2} \qquad \text{Gaussian well type}$$

$$V(r) = -V_0\ \mathrm{e}^{-(r/a)} \qquad \text{Exponential well type}$$

$$V(r) = -V_0\ \frac{\mathrm{e}^{-r/a}}{r} \qquad \text{Yukawa well type}$$

are used by various scientists and each of them also produced very similar type of results.

For the ground state of the deuteron, the total energy E is negative and equal to $-B$, where B is the binding energy of deuteron. Thus equation (2.13) can be written as

$$\frac{\mathrm{d}^2u}{\mathrm{d}r^2} + \frac{m}{\hbar^2}\Big[V_0 - B\Big]u = 0 \qquad \text{for} \qquad r < b \tag{2.14}$$

and

$$\frac{\mathrm{d}^2u}{\mathrm{d}r^2} - \frac{mB}{\hbar^2}\ u = 0 \qquad \text{for} \qquad r > b \tag{2.15}$$

These equations can be written as

$$\frac{\mathrm{d}^2u}{\mathrm{d}r^2} + K^2u = 0 \qquad \text{for} \qquad r < b \tag{2.16}$$

and

$$\frac{d^2u}{dr^2} - \alpha^2 u = 0 \qquad \text{for} \qquad r > b \tag{2.17}$$

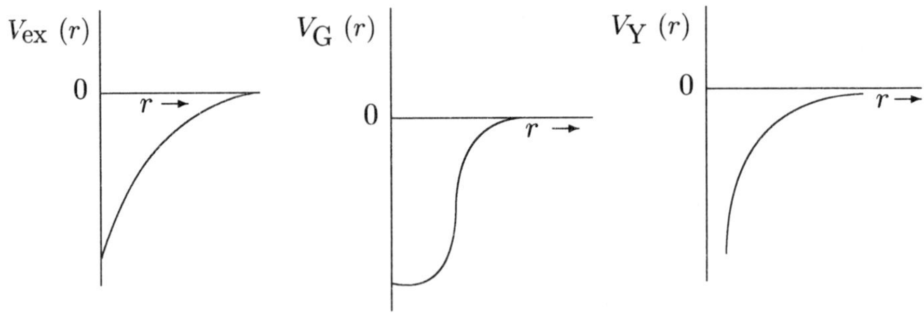

(a) Exponential (b) Gaussian (c) Yukawa

Figure 4: Some other forms used by various scientists for the potential of deuteron.

where $K^2 = m(V_0 - B)/\hbar^2$ and $\alpha^2 = mB/\hbar^2$. General solutions of equations (2.16) and (2.17), respectively, are

$$u = A_1 \sin Kr + B_1 \cos Kr \qquad \text{for} \qquad r < b \tag{2.18}$$

and

$$u = A_2 \, e^{\alpha r} + B_2 \, e^{-\alpha r} \qquad \text{for} \qquad r > b \tag{2.19}$$

On these solutions, the boundary conditions need to be imposed are as the following.

1. For $r \to 0$, we must have $u \to 0$, so that the wave function R is finite.

2. For $r \to \infty$, we must have $u \to 0$, as the particle has zero probability to be found at infinity.

To satisfy the conditions at $r = 0$, we get $B_1 = 0$ and to satisfy the conditions at $r = \infty$, we get $A_2 = 0$. Thus, equations (2.18) and (2.19), respectively, reduce to

$$u = A_1 \sin Kr \qquad \text{for} \qquad r < b \tag{2.20}$$

and

$$u = B_2 \, e^{-\alpha r} \qquad \text{for} \qquad r > b \tag{2.21}$$

Since these two solutions join smoothly at $r = b$. Hence, equating the wave functions u and the first derivatives of u at $r = b$, we, respectively, have

$$A_1 \sin Kb = B_2 \, e^{-\alpha b} \tag{2.22}$$

and

$$A_1 K \cos Kb = -B_2 \, \alpha \, e^{-\alpha b} \tag{2.23}$$

Therefore, we have

$$K \cot Kb = -\alpha \tag{2.24}$$

Using the values of K and α, we get

$$\cot \left[\frac{b\sqrt{m(V_0 - B)}}{\hbar}\right] = -\left(\frac{B}{V_0 - B}\right)^{1/2}$$

or

$$\tan \left[\pi - \frac{b\sqrt{m(V_0 - B)}}{\hbar}\right] = \left(\frac{V_0 - B}{B}\right)^{1/2} \tag{2.25}$$

Thus, for a given value of binding energy B, equation (2.25) is a relation between b and V_0. This transcendental equation (2.25) can be solved numerically as well as graphically. For the binding energy $B = 2.225$ MeV, the values of the range b corresponding to the potential V_0 can be obtained, as given in Table 1.

Table 1: Values of b corresponding to V_0 for $B = 2.225$ MeV

V_0 (MeV) =	110	100	90	70	60	40	30	20
b (10^{-15} m) =	1.04	1.10	1.16	1.34	1.47	1.86	2.21	2.86

Several experiments have given a approximate range of the nuclear force to be of the order of 10^{-15}m. For the range $b \approx 2 \times 10^{-15}$ m, the depth V_0 should be about 36 MeV (Table 1).

3.2 Probable S states of deuteron

We already know that the ground state of deuteron is an ^{3}S state. Though there is no experimental evidence about the existence of any excited state of deuteron. However, let us look about it from the theory point of view. First, we shall see about the S states and then about the P states.

Let us see if some excited S state of deuteron can be found theoretically. Equation (2.24) can be written as

$$x \; \cot x = -\alpha b \tag{2.26}$$

where $x = Kb$. Using $B = 2.225$ MeV, the value of $\alpha = 0.237 \times 10^{15}$ m^{-1}. For $b \leq 3 \times 10^{-15}$ m, we get $\alpha b \leq 0.7$. Roots of equation (**??**) can be obtained graphically by plotting the curves $y = \cot x$ and $y = -\alpha b/x$. The intersections of theses curves give the roots. These two graphs are plotted in Figure 5.

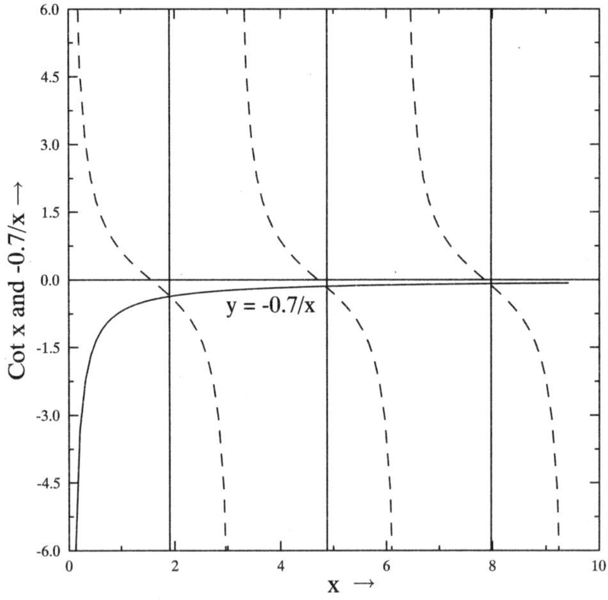

Figure 5: Graphs of $y = \cot x$ and $y = -0.7/x$.

The figure shows that the roots are slightly greater than $(2n+1)\pi/2$, where $n = 0, 1, 2, \ldots$. Let the roots be $x_0 = (2n+1)\pi/2 + \epsilon$. Thus, we have

$$\left[\frac{(2n+1)\pi}{2} + \epsilon\right] \cot\left[\frac{(2n+1)\pi}{2} + \epsilon\right] = -\alpha b \tag{2.27}$$

Since ϵ is small, we have $\cot[(2n+1)\pi/2 + \epsilon] = -\epsilon$ and thus, we get

$$\frac{(2n+1)\pi}{2}\epsilon = \alpha b \qquad \text{or} \qquad \epsilon = \frac{2\alpha b}{(2n+1)\pi} \tag{2.28}$$

Thus, using $x_0 = Kb$, we have

$$\frac{b\sqrt{m(V_0 - B)}}{\hbar} = \frac{(2n+1)\pi}{2} + \frac{2\alpha b}{(2n+1)\pi}$$

or

$$V_0 = B + \left[\frac{(2n+1)\pi}{2} + \frac{2\alpha b}{(2n+1)\pi}\right]^2 \frac{\hbar^2}{b^2 m} \tag{2.29}$$

For $B = 2.225$ MeV, the values of V_0 for some values of b and n are calculated and given in Table 2.

Table 2: Values of V_0 in MeV.

b (10^{-15} m)	$n = 0$	$n = 1$	$n = 2$
1.5	59.39	408. 00	1106.76
2.0	37.14	232.88	625.90
2.5	26.41	151.39	402.90
3.0	20.34	106.89	231.53

For $V_0 >> B$ and neglecting the second term in comparison to the first one in the square bracket, we have[2]

$$V_0 b^2 = \frac{h^2}{16m}, \frac{9h^2}{16m}, \frac{25h^2}{16m}, \ldots$$

Here, the first term on right side, corresponds to the ground state. The second, third, ... terms corresponds to different excited S states. It is found that

$$V_0 b^2 = \frac{9h^2}{16m}, \frac{25h^2}{16m}, \ldots$$

corresponds to a node inside the square-well potential. Thus, we cannot have any excited S state. For the ground state, we have

$$V_0 b^2 = \frac{h^2}{16m}$$

For $b = 2 \times 10^{-15}$ m, we have

$$V_0 = \frac{h^2}{16mb^2}$$

$$= \frac{(6.626 \times 10^{-34})^2}{16 \times 1.67 \times 10^{-27} \times (2 \times 10^{-15})^2} = 4.11 \times 10^{-12}\,\text{J}$$

$$= 24.6 \text{ MeV}$$

[2]In an alternative approach, we can obtain this relation as the following. When $V_0 >> B$, equation (2.25) gives

$$\cot\left[\frac{b\sqrt{mV_0}}{\hbar}\right] = 0 \qquad \text{or} \qquad \frac{b\sqrt{mV_0}}{\hbar} = \frac{\pi}{2}, \frac{3\pi}{2}, \frac{5\pi}{2}$$

It gives

$$V_0 b^2 = \frac{h^2}{16m}, \frac{9h^2}{16m}, \frac{25h^2}{16m}, \ldots$$

3.3 Normalization of wave function

The constants A_1 and B_2 are obtained from the requirement that the integral of $|\psi|^2$ over all the space must be equal to unity.

$$4\pi \int_0^\infty |R|^2 r^2 \mathrm{d}r = 4\pi \int_0^\infty u^2 \mathrm{d}r = 1$$

or

$$4\pi \int_0^b A_1^2 \sin^2 Kr \, \mathrm{d}r + 4\pi \int_b^\infty B_2^2 \, \mathrm{e}^{-2\alpha r} \mathrm{d}r = 1$$

It gives

$$A_1^2 \left[b - \frac{1}{2K} \sin 2Kb \right] + (B_2^2/\alpha) \, \mathrm{e}^{-2\alpha b} = \frac{1}{2\pi} \tag{2.30}$$

With the help of equations (2.22) and (2.30), we get

$$A_1 = \left[\frac{\alpha}{2\pi(1 + b\alpha)} \right]^{1/2} \qquad \text{and} \qquad B_2 = \left[\frac{\alpha}{2\pi(1 + b\alpha)} \right]^{1/2} \mathrm{e}^{\alpha b} \sin Kb$$

Thus, the radial wave functions for the deuteron are

$$u = \left[\frac{\alpha}{2\pi(1 + b\alpha)} \right]^{1/2} \sin Kr \qquad \text{for} \qquad r < b$$

and

$$u = \left[\frac{\alpha}{2\pi(1 + b\alpha)} \right]^{1/2} \mathrm{e}^{\alpha(b-r)} \sin Kb \qquad \text{for} \qquad r > b$$

Figure 6 shows variation of $u(r)$ is a function of r.

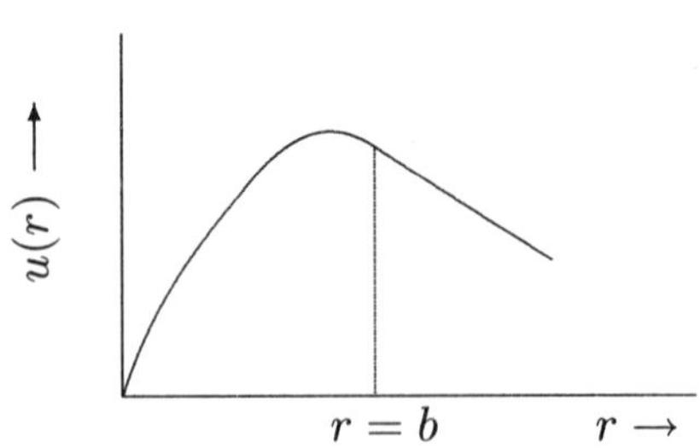

Figure 6: Ground state wave function of deuteron.

The probability of finding the nucleons within the range of the nuclear potential is

$$P = \frac{4\pi\alpha}{2\pi(1 + b\alpha)} \int_0^b \sin^2 Kr \, \mathrm{d}r = \frac{2\alpha}{(1 + b\alpha)} \frac{b}{2} = \frac{b\alpha}{(1 + b\alpha)}$$

For $m = 1.67 \times 10^{-27}$ kg, $B = 2.225$ MeV, $b = 2 \times 10^{-15}$ m, we get $P = 0.315$ or 31.5%

Thus, the probability of the nucleons within the range of nuclear force in the deuteron is about 32%. Hence, the probabilty of finding the neucleons away from the range of the nuclear force is twice that of within the range. It shows that the deuteron is a loosely bound structure (low binding energy, 1.113 MeV per nucleon). From the classical point of view, the nucleons cannot stay bound outside the potential well. But, quantum mechanically we find that the nucleons can stay outside the well.

3.4 Size of deuteron

The distance r_d at which the exponential function (for $r > b$) falls to 1/e of its initial value. Thus, we should have

$$\alpha r_d = 1 \qquad \text{therefore} \qquad r_d = \frac{1}{\alpha}$$

Now, for $m = 1.67 \times 10^{-27}$ kg, $B = 2.225$ MeV, we have

$$\alpha = \frac{\sqrt{mB}}{\hbar} = 0.237 \times 10^{15} \text{ m}$$

Thus, the radius of deuteron is

$$r_d = \frac{1}{\alpha} = 4.2 \times 10^{-15} = 4.2 \text{ F}$$

It shows that the radius of deuteron is twice the range of nuclear potential. This supports the fact that deuteron is a loosely bound structure.

3.5 Magnetic moment of deuteron

The magnetic moment of deuteron is due to the orbital motion of proton having positive electric charge and due to the intrinsic spin of each of nucleon. Because of zero electric charge, neutron has no contribution due to its orbital motion. The magnetic moment μ_l arising due to orbital motion, in the units of nuclear magneton μ_N is

$$\mu_l = g_l \, l$$

where l is the total angular momentum quantum number of a nucleon and g_l is 1 for proton and zero for neutron.

3.6 Excited P state of deuteron

The ground state of deuteron with the lowest energy corresponds to the orbital angular momentum quantum number $l = 0$. We have seen that deuteron does not have any excited S state. Let us now look into the matter if deuteron has any excited P state. The radial part of Schrödinger for $l = 1$ (P state) can be expressed as

$$\frac{1}{r^2}\frac{\mathrm{d}}{\mathrm{d}r}\left(r^2\frac{\mathrm{d}R}{\mathrm{d}r}\right) + \frac{2\mu}{\hbar^2}\left[E - V(r) - \frac{\hbar^2}{\mu r^2}\right]R = 0 \qquad (2.31)$$

For the deuteron, the reduced mass $\mu = m/2$ and substituting $R = u(r)/r$ in equation (2.31), where $u(r)$ is called the radial wave function, we have

$$\frac{\mathrm{d}^2u_l(r)}{\mathrm{d}r^2} + \left[K^2 - \frac{2}{r}\right]u_l(r) = 0 \qquad r < b \qquad (2.32)$$

and

$$\frac{\mathrm{d}^2u_l(r)}{\mathrm{d}r^2} - \left[\alpha^2 + \frac{2}{r}\right]u_l(r) = 0 \qquad r > b \qquad (2.33)$$

where $K^2 = m(V_0 - B)/\hbar^2$ and $\alpha^2 = mB/\hbar^2$. For the limit $B \to 0$, solutions of equations (2.32) and (2.33) can be obtained in terms of spherical Bessel functions and spherical Neumann functions. Using boundary conditions that the function and its first derivative are continuous at $r = b$, we get that Kb should be equal to π. That is,

$$\frac{b\sqrt{m(V_0 - B)}}{\hbar} = \pi$$

For $B \to 0$, we have

$$V_0 = \frac{h^2}{4mb^2}$$

If we choose $b = 2 \times 10^{-15}$m, we get $V_0 = 144$ MeV, which is almost four times as large as the actual well depth in the ground state. It shows that there is no excited P state.

Repeating this procedure for large values of l, we find that a deeper and deeper well depth is required to produce a bound state. Thus we conclude that no bound state exists for $l > 0$.

4. Nucleon-nucleon scattering

With the help of the binding energy of deuteron, obtained experimentally, we could derive important information that the force between a proton and a neutron in the deuteron is attractive and of short range. Deuteron has no excited state and in the found state, the spins of proton and neutron are parallel and it cannot provide information about the mutual interaction between a proton and a neutron. For the force of attraction, the two spins should be anti-parallel. One would also like to know the interaction between a pair of protons and between a pair of neutrons. One can learn more about the interaction between two nucleons by shooting one nucleon towards the other. For a short duration, the two nucleons arrive within the range of nuclear force and then they move away from each other. In this process, the projectile is scattered from its original path. With the help of such scattering experiments, we can learn about the forces, which are responsible for the scattering.

In order to describe the scattering theoretically, it is convenient to use the coordinates where the centre-of-mass of the system (incident particle plus the target particle) is at rest, in both the situations, before and after the scattering. Such a system is known as the centre-of-mass coordinate system. In this system obviously the total momentum is always zero. For this reason, the centre-of-mass system is sometimes known as the zero-momentum system. On the other side, the laboratory observations are made where the target is kept initially at rest. This system used for the observations is known as the lab-system. In order to compare theory and experiment, it would be interesting to have information about the transformation of certain relevant physical quantities from one system into the other. Let us first describe these two coordinate systems.

4.1 Laboratory system

Suppose a particle (called projectile) of mass m_1 moves with initial velocity u towards a stationary target of mass m_2. As the mass m_1 moves towards m_2, the centre-of-mass moves with a velocity v_{cm} along the direction of the incident particle. After the elastic collision, let the incident particle moves with velocity v_1 making an angle θ_L with the initial direction and the target particle moves with velocity v_2 making an angle ϕ_L with the initial direction (Figure 7).

For the velocity v_{cm} of the centre-of-mass before collision, the conservation of angular momentum gives

$$m_1 u = (m_1 + m_2) v_{cm} \qquad \text{or} \qquad v_{cm} = \frac{m_1 u}{(m_1 + m_2)} \tag{2.34}$$

After collision, the conservation of angular momentum along the initial direction gives

$$m_1 u = m_1 v_1 \cos\theta_L + m_2 v_2 \cos\phi_L \tag{2.35}$$

and the conservation of angular momentum perpendicular to the initial direction gives

$$0 = m_1 v_1 \sin\theta_L - m_2 v_2 \sin\phi_L \tag{2.36}$$

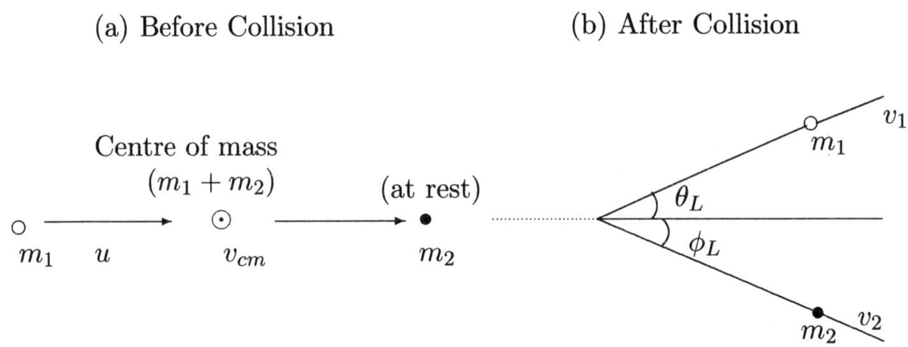

Figure 7: Collision of two particles in the centre-of-mass system.

For the elastic collision, the conservation of energy gives

$$\frac{1}{2}\, m_1 u^2 = \frac{1}{2}\, m_1 v_1^2 + \frac{1}{2}\, m_2 v_2^2 \tag{2.37}$$

These three equations (2.35), (2.36) and (2.37), relate four quantities v_1, v_2, θ_L and ϕ_L as a function of the initial velocity u. Here, we have four unknowns and three equations. When one of the v_1, v_2, θ_L and ϕ_L, values of the rest three can be obtained.

4.2 Center-of-mass system

In the center-of-mass system, both the incident particle as well as the target move towards the center-of-mass in such a manner that the center-of-mass remains stationary. Both the incident particle as well as the target reach the center-of-mass simultaneously and interact. After interaction, both of them move away from the center-of-mass in the directions opposite to each other and making an angle θ_c, as shown in Figure 8. Before collision, let u_1 and u_2 be velocities of the incident particle and target, respectively, relative to the center-of-mass. Thus, we have

$$u_1 = u - v_{cm} = u - \frac{m_1 u}{(m_1 + m_2)} = \frac{m_2 u}{(m_1 + m_2)}$$

and

$$u_2 = 0 - v_{cm} = -\frac{m_1 u}{(m_1 + m_2)}$$

Momenta of particles along the initial direction before collision are

$$P_1 = m_1u_1 = \frac{m_1m_2u}{(m_1+m_2)} = \mu u$$

and

$$P_2 = m_2u_2 = -\frac{m_1m_2u}{(m_1+m_2)} - \mu u$$

where $\mu = m_1m_2/(m_1+m_2)$ is the reduced-mass of the system. Thus, the two particles have equal and opposite momentum before collision ($P_1 = -P_2$), and therefore total momentum is zero. The conservation of momentum thus requires that after the collision, the total momentum of the system must be zero. If P_1' and P_2' are momenta of particles after collision, we have $P_1' = -P_2'$. Since there is no interaction between the two particles, before as well as after the collision, the potential energy is zero. Now, the conservation of energy requires

$$E_1 + E_2 = E_1' + E_2' \tag{2.38}$$

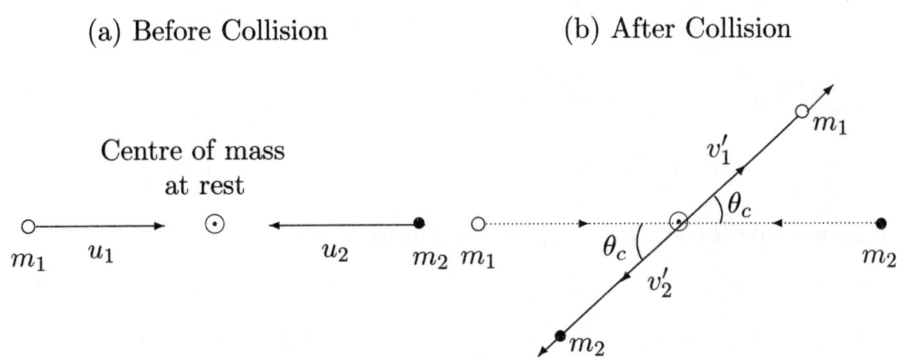

Figure 8: Collision of two particles in the lab-system.

where the kinetic energies E_1, E_2, E_1' and E_2' are

$$E_1 = \frac{P_1^2}{2m_1}, \qquad E_2 = \frac{P_2^2}{2m_2}, \qquad E_1' = \frac{P_1'^2}{2m_1}, \qquad E_2' = \frac{P_2'^2}{2m_2} \tag{2.39}$$

On using equation (2.39) in (2.38), we get

$$\frac{P_1^2}{2m_1} + \frac{P_2^2}{2m_2} = \frac{P_1'^2}{2m_1} + \frac{P_2'^2}{2m_2} \tag{2.40}$$

Since $P_1 = -P_2$ and $P_1' = -P_2'$, equation (2.40) gives

$$P_1^2\left[\frac{1}{2m_1} + \frac{1}{2m_2}\right] = P_1'^2\left[\frac{1}{2m_1} + \frac{1}{2m_2}\right] \tag{2.41}$$

It gives

$$P_1 = P_1' \qquad \text{or} \qquad u_1 = v_1'$$

and similarly, we can have

$$P_2 = P_2' \qquad \text{or} \qquad u_2 = v_2'$$

Thus, in the centre-of-mass system, the velocities of particles remain unchanged. Only the directions of the particles change.

4.3 Relation between θ_L and θ_c

The velocity v_1 of particle m_1 after scattering in the lab-system is obtained by adding vectorially the velocity v_{cm} of the centre-of-mass in lab-system to v_1' of m_1 in the centre-of-mass system after scattering. This vector relationship between θ_L and θ_c is shown in Figure 8.

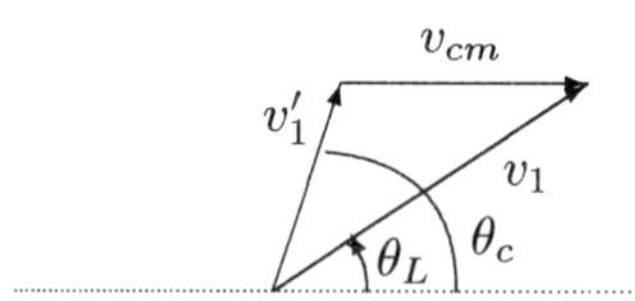

Figure 9: Vector relationship between θ_L and θ_c.

Thus, we have $\vec{v_1} = \vec{v_1}' + \vec{v_{cm}}$. From Figure 9, we have

$$v_1 \cos\theta_L = v_1' \cos\theta_c + v_{cm} \qquad \text{and} \qquad v_1 \sin\theta_L = v_1' \sin\theta_c \tag{2.42}$$

The scattering being in the same plane, the azimuthal angle in the lab-system as well as in the centre-of-mass system remain the same. Thus, from equations (2.42), we get

$$\tan\theta_L = \frac{v_1' \sin\theta_c}{v_1' \cos\theta_c + v_{cm}} = \frac{\sin\theta_c}{\cos\theta_c + \gamma} \tag{2.43}$$

where $\gamma = v_{cm}/v_1'$. This equation gives relation between θ_L and θ_c. From the relations

$$v_{cm} = \frac{m_1 u}{(m_1 + m_2)} \qquad \text{and} \qquad u_1 = \frac{m_2 u}{(m_1 + m_2)}$$

we get

$$\frac{v_{cm}}{u_1} = \frac{m_1}{m_2}$$

Since $u_1 = v_1'$, we have $\gamma = m_1/m_2$.

Special cases for elastic collisions

Let us discuss some special cases for the relation between θ_L and θ_c.

(i) When the target is much heavier than the projectile ($m_2 >> m_1$), then the value of γ is very small than can be neglected. Therefore $\tan\theta_L = \tan\theta_c$ or $\theta_L = \theta_c$.

(ii) When the target is much lighter than the projectile ($m_2 << m_1$), then the value of γ is very large. Therefore $\tan\theta_L \approx \infty$ or $\theta_L = \pi/2$.

(iii) When the mass of target is equal to that of the projectile ($m_1 = m_2$), then we have

$$\tan\theta_L = \frac{\sin\theta_c}{1+\cos\theta_c} \qquad \text{or} \qquad \theta_L = \frac{\theta_c}{2}$$

4.4 Inelastic scattering

So far we have discussed about the elastic scattering, where there was no exchange of energy. Projectile as well as the target were retaining their kinetic energies and only their paths were changing. When the scattering is such that some internal energy of the system is converted into the kinetic energy of the scattered particles, then the scattering is known as the inelastic scattering. The momentum conservation condition in the center-of-mass system remains the same, as

$$P_1 + P_2 = P_1' + P_2' = 0$$

However, the energy conservation is now as the following.

$$\frac{P_1^2}{2m_1} + \frac{P_2^2}{2m_2} = \frac{P_1'^2}{2m_1} + \frac{P_2'^2}{2m_2} + Q \qquad (2.44)$$

where Q is the amount of energy which is lost or gained as a result of impact. In the nuclear reactions, Q is of fundamental importance. When Q is positive, the reaction is said to be exothermic and when Q is negative, the reaction is said to be endothermic. For this case also, equations (2.42) and (2.43) are valid and γ is still defined as v_{cm}/v_1'. Now, the value of γ becomes

$$\gamma = \frac{m_1}{m_2}\left[1 - \frac{Q}{E}\left(1 + \frac{m_1}{m_2}\right)\right]^{-1/2}$$

where E is the kinetic energy of the incident particle as measured in the lab-system.

5. Scattering cross sections

In an experiments on scattering, the projectile is scattered from its initial direction under the influence of the field generated by the target. The probability of scattering is expressed in terms of a quantity, known as the scattering cross-section. The cross-section represents the probability of scattering of projectile from its initial path. It also provides information about the number of particles scattered from the initial beam of projectile particles, after interaction with the target. The angular distribution of particles scattered is, however, determined by differential scattering cross-section.

Suppose a parallel beam of particles having n particles per unit area impinges on a target. Due to interaction between the incident particles and the target, the particles are scattered in all directions. The number of particles which are scattered per unit time through an element of solid angle $\mathrm{d}\Omega$ about the direction making an angle θ with respect to the direction of incidence are proportional to n and $\mathrm{d}\Omega$ (Figure 10). Here, we are assuming that (i) the incoming flux n is so small that there is no interaction among the incident particles, *i.e.*, the particles are moving in an independent manner. (ii) The separation between various targets are quite large, so that participation of individual target can be considered.

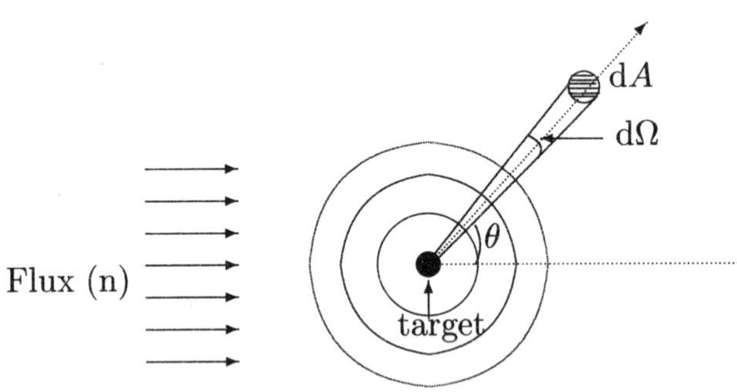

Figure 10: Scattering Process

Let $n_s(\theta)$ be the number of particles scattered in a solid angle $\mathrm{d}\Omega$ per sec about the direction θ. Then, we have

$$n_s(\theta) \propto n\ \mathrm{d}\Omega \qquad \text{or} \qquad n_s(\theta) = \sigma(\theta) n\ \mathrm{d}\Omega \tag{2.45}$$

where $\sigma(\theta)$ is known as the differential scattering cross section. The word, cross section is used as the dimensions of $\sigma(\theta)$ are of an area. Thus, we have

$$\sigma(\theta) = \frac{n_s(\theta)}{n\ \mathrm{d}\Omega}$$

Here, it is assumed that the scattering is independent of azimuthal angle ϕ with respect to the direction of incidence. For an unpolarized beam of incident particles and for the spherically symmetric potential, the scattering is independent of the azimuthal angle ϕ . The number of particles scattered by a target, into the solid angle $\mathrm{d}\Omega$ per unit flux of incident particles is given as $\sigma(\theta)\ \mathrm{d}\Omega$.

The total scattering cross section is obtained by integrating the differential scattering cross section over all directions as

$$\sigma = \int \sigma(\theta)\ \mathrm{d}\Omega$$

As $\sigma(\theta)$ is independent of ϕ, we have

$$\sigma = \int_0^{2\pi} \mathrm{d}\phi \int_0^{\pi} \sigma(\theta) \sin\theta\ \mathrm{d}\theta = 2\pi \int_0^{\pi} \sigma(\theta) \sin\theta\ \mathrm{d}\theta \tag{2.46}$$

The scattering cross section is thus defined by the total number of particles scattered in all directions. It is expressed in barns, where 1 barn $= 10^{-28}$ m^2. The differential scattering cross section is expressed in barns/str.

5.1 Relation between scattering cross sections

When the scatterer (target) is heavy enough (having very large mass), the center of masses (projectile + target) coincides with center of the target and the center-of-mass always remains at rest. The equation (2.45) is valid in the both the lab as well as the center-of-mass coordinate systems. Consequently, $\sigma(\theta)$ expressed by equation (2.46) remains the same in both the systems. However, when the target is light in mass, the center of masses does not coincide with center of the target, and it moves. Consequently, equation (2.45) is applicable to the lab-system only. Since the theory is developed in the center-of-mass system whereas the experiments are performed in the lab-system, it is therefore required to have some laws of transformation of cross section from one coordinate system to other.

By definition of scattering cross section, it is required that the number of particles scattered into an element of solid angle $\mathrm{d}\Omega_L$ about the direction θ_L in lab-system should be the same as scattered into an element of solid angle $\mathrm{d}\Omega_c$ in the center-of-mass system. Thus, we have

$$n\sigma_L(\theta_L)\ \mathrm{d}\Omega_L = n\sigma_c(\theta_c)\ \mathrm{d}\Omega_c \tag{2.47}$$

where $\sigma_L(\theta_L)$ and $\sigma_c(\theta_c)$ are the scattering cross sections in the two systems. As the scattering is independent of azimuthal angle ϕ, we have

$\phi_L = \phi_c$ and therefore, $\mathrm{d}\phi_L = \mathrm{d}\phi_c$. From equation (2.47), we get

$$\sigma_L(\theta_L) 2\pi \sin\theta_L \, \mathrm{d}\theta_L = \sigma_c(\theta_c) 2\pi \sin\theta_c \, \mathrm{d}\theta_c$$

or

$$\sigma_L(\theta_L) = \frac{\sin\theta_c \, \mathrm{d}\theta_c}{\sin\theta_L \, \mathrm{d}\theta_L} \sigma_c(\theta_c) \tag{2.48}$$

On differentiating equation (2.43), we get

$$\sec^2\theta_L \, \mathrm{d}\theta_L = \frac{(1+\gamma\cos\theta_c)}{(\gamma+\cos\theta_c)^2} \, \mathrm{d}\theta_c$$

or

$$\frac{\mathrm{d}\theta_c}{\mathrm{d}\theta_L} = \frac{(\gamma+\cos\theta_c)^2}{(1+\gamma\cos\theta_c)\cos^2\theta_L} \tag{2.49}$$

Further, from equation (2.43), we get

$$\sec^2\theta_L = \tan^2\theta_L + 1 = \left(\frac{\sin\theta_c}{\gamma+\cos\theta_c}\right)^2 + 1 = \frac{1+\gamma^2+2\gamma\cos\theta_c}{(\gamma+\cos\theta_c)^2}$$

Therefore, we have

$$\cos^2\theta_L = \frac{(\gamma+\cos\theta_c)^2}{(1+\gamma^2+2\gamma\cos\theta_c)} \tag{2.50}$$

Now, we have

$$\sin\theta_L = \sqrt{1-\cos^2\theta_L} = \frac{\sin\theta_c}{(1+\gamma^2+2\gamma\cos\theta_c)^{1/2}} \tag{2.51}$$

Using equations (2.49), (2.50) and (2.51) in (2.48), we get

$$\sigma_L(\theta_L) = \frac{(1+\gamma^2+2\gamma\cos\theta_c)^{3/2}}{1+\gamma\cos\theta_c} \sigma_c(\theta_c) \tag{2.52}$$

This is the required relation between the scattering cross sections in the two systems. However, the total scattering cross section remains the same in the two systems. When $m_1 = m_2$ and the scattering is elastic, then we have $\gamma = 1$, $\theta_L = \theta_c/2$ and equation (2.52) gives

$$\sigma_L(\theta_L) = 4\cos\theta_L \; \sigma_c(\theta_c)$$

5.2 Relation between the cross section and amplitude for the elastic scattering

In an experiment for scattering, a collimated beam of incident particles moving with velocity v along the direction of $+z$-axis, say, is allowed to incident on a target, situated at the origin of the coordinate system. The beam of particles having a plane wavefront can be represented as

$$\psi_{inc} = \mathrm{e}^{(ikz-i\omega t)}$$

where the wave number $k = 2\pi/\lambda = \sqrt{2\mu E}/\hbar$ and ω is the angular frequency. Here, λ represents the de Broglie wavelength for a particle of reduced mass μ and energy E in the center-of-mass system. The amplitude of the wave is taken to be unity, so that the wave represents a particle density of one particle per unit volume. Thus, the number of particles flowing through a unit cross-sectional area in a unit time is v. In other words, the incident flux of particles is just equal to v. In the case of a continuous sequence of interactions between the projectiles and the target, the situation is like a steady state and then one can take away the time factor in the incident plane wave. Thus, we have

$$\psi_{inc} = \mathrm{e}^{ikz} \tag{2.53}$$

The probability current density is

$$S_{inc} = \frac{\hbar}{2i\mu}\left[\psi^*\nabla\psi - \psi\nabla\psi^*\right] \tag{2.54}$$

Using equation (2.53) in (2.54), we get

$$S_{inc} = \frac{\hbar}{2i\mu}\left[\mathrm{e}^{-ikz}\frac{\mathrm{d}}{\mathrm{d}z}\,\mathrm{e}^{ikz} - \mathrm{e}^{ikz}\frac{\mathrm{d}}{\mathrm{d}z}\,\mathrm{e}^{-ikz}\right] = \frac{\hbar k}{\mu} = \frac{p}{\mu} = v$$

Thus, at vary large distances from the target, the incident flux is v, *i.e.*, equal to the velocity of particles at these distances.

During interaction, the incident particles are scattered from their path. After that the scattered particles go away from the target and are described by a spherical wave having axial symmetry radiating outward from the center of interaction (Figure 10). At large distances from the target, the wave function is a linear combination of the incident beam (not scattered by the target) and a spherical wave representing the scattered particles, as

$$\psi = \psi_{inc} + \psi_{sc} \qquad \text{or} \qquad \psi = \mathrm{e}^{ikz} + f(\theta)\,\frac{\mathrm{e}^{ikr}}{r} \tag{2.55}$$

where $f(\theta)$ is known as the scattering amplitude for the elastic scattering. The particle density in the incident beam is so small that there is no interference between the incident and scattered particles. The scattering amplitude is independent of azimuthal angle ϕ. That is, the scattering is symmetric about the direction of incident beam. Here, $f(\theta)$ is a measure of scattered wave in the direction having polar angle θ. Since the amplitude of spherical wave fall off as $1/r$ and therefore the radial flux as $1/r^2$, we can neglect the terms higher than $1/r^2$ in the estimation of the probability current density for the scattered spherical wave. For the scattered spherical wave, the current density is

$$S_{sc} = \frac{\hbar}{2i\mu}\left[\psi^*\nabla\psi - \psi\nabla\psi^*\right]$$

$$= \frac{\hbar}{2i\mu}\left[f(\theta)^* \frac{\mathrm{e}^{-ikr}}{r}\frac{\mathrm{d}}{\mathrm{d}r}\left(f(\theta)\frac{\mathrm{e}^{ikr}}{r}\right) - f(\theta)\frac{\mathrm{e}^{ikr}}{r}\frac{\mathrm{d}}{\mathrm{d}r}\left(f(\theta)^*\frac{\mathrm{e}^{-ikr}}{r}\right)\right]$$

$$= |f(\theta)|^2\frac{\hbar k}{\mu r^2} = \frac{v}{r^2}|f(\theta)|^2 \tag{2.56}$$

The number of particles in the scattered wave which cross an element of area $\mathrm{d}A$ perpendicular to radius vector in unit time (Figure 10) is

$$S_{sc}\,\mathrm{d}A = \frac{v}{r^2}|f(\theta)|^2\mathrm{d}A = v|f(\theta)|^2\mathrm{d}\Omega$$

where $\mathrm{d}\Omega = \mathrm{d}A/r^2$ is the solid angle subtended by the area $\mathrm{d}A$ at the origin (target). As per definition, the differential cross section $\sigma(\theta)$ is the ratio of number of particles scattered per unit solid angle $\mathrm{d}\Omega$ at an angle θ to the incident flux and therefore, we have

$$\sigma(\theta) = \frac{v|f(\theta)|^2}{v} = |f(\theta)|^2$$

This is the relation between the differential cross section $\sigma(\theta)$ and the amplitude $f(\theta)$ of the scattered wave. The total cross section σ for the scattering is obtained by integrating the differential cross section over all directions, as

$$\sigma = \int\sigma(\theta)\,\mathrm{d}\Omega = \int\sigma(\theta)\sin\theta\,\mathrm{d}\theta\,\mathrm{d}\phi = 2\pi\int_0^{\pi}|f(\theta)|^2\sin\theta\,\mathrm{d}\theta \tag{2.57}$$

5.3 Partial wave analysis

This technique of partial wave analysis is suitable for low energy scattering and a spherically symmetric potential of short range. The incident plane wave can be expressed as

$$\psi_{inc} = e^{ikz} = e^{ikr\cos\theta}$$

This expression can be expanded as

$$\psi_{inc} = e^{ikr\cos\theta} = \sum_{l=0}^{\infty}(2l+1)i^l j_l(kr)P_l(\cos\theta)$$

where $P_l(\cos\theta)$ is the Legendre polynomial and $j_l(kr)$ the Bessel function. The terms corresponding to various values of l are known as the partial waves. In the scattering experiment, the particles are detected at large distance from the target, for $r \to \infty$, we have

$$j_l(kr) = \frac{\sin(kr - l\pi/2)}{kr}$$

Therefore, the wave function at large distance is

$$\psi_{inc} = \sum_{l=0}^{\infty}(2l+1)i^l P_l(\cos\theta)\,\frac{\sin(kr - l\pi/2)}{kr} \tag{2.58}$$

or

$$\psi_{inc} = \frac{1}{kr}\sum_{l=0}^{\infty}(2l+1)i^l P_l(\cos\theta)\,\frac{e^{i(kr-l\pi/2)} - e^{-i(kr-l\pi/2)}}{2i} \tag{2.59}$$

This asymptotic form of the incident plane wave, expressed by equation (2.59), can be considered as a series of outgoing spherical waves $e^{i(kr-l\pi/2)}$ along with a series of incoming spherical waves $e^{-i(kr-l\pi/2)}$ of equal amplitudes. The presence of scattering potential affects the outgoing waves both in amplitude and in phase. The incoming waves remain unaffected. The asymptotic form of total wave function can be expressed as

$$\psi = \frac{1}{kr}\sum_{l=0}^{\infty}(2l+1)i^l P_l(\cos\theta)\,\frac{\eta_l\, e^{i(kr-l\pi/2)} - e^{-i(kr-l\pi/2)}}{2i} \tag{2.60}$$

Here, the factor η_l represents the effect of interaction potential on the outgoing waves. It is in general a complex number. When η_l is real or $|\eta_l| = 1$, the phase of the outgoing wave changes and it corresponds to elastic scattering. When η_l is imaginary or $|\eta_l| < 1$, both phase and amplitude are affected, which then correspond to absorption of particles. As we are interested only in the elastic scattering, we shall consider only $|\eta_l| = 1$.

The total wave function given by (2.60) can be considered as superposition of incident and scattered waves

$$\psi_{r\to\infty} = \psi_{inc} + \psi_{sc} \tag{2.61}$$

According to equation (2.55), we know

$$\psi_{sc} = f(\theta)\ \frac{e^{ikr}}{r} \tag{2.62}$$

However, from equation (2.61, we have

$$\psi_{sc} = \psi_{r\to\infty} - \psi_{inc}$$

$$= \frac{1}{kr}\sum_{l=0}^{\infty}(2l+1)i^l P_l(\cos\theta)\ \frac{\eta_l\ e^{i(kr-l\pi/2)} - e^{-i(kr-l\pi/2)}}{2i}$$

$$-\frac{1}{kr}\sum_{l=0}^{\infty}(2l+1)i^l P_l(\cos\theta)\ \frac{e^{i(kr-l\pi/2)} - e^{-i(kr-l\pi/2)}}{2i}$$

$$= \frac{1}{kr}\sum_{l=0}^{\infty}(2l+1)i^l P_l(\cos\theta)\ \frac{e^{i(kr-l\pi/2)}}{2i}(\eta_l - 1)$$

$$= \frac{1}{2ik}\sum_{l=0}^{\infty}(2l+1)i^l P_l(\cos\theta)\ \frac{e^{ikr}}{r}\ e^{-il\pi/2}(\eta_l - 1) \tag{2.63}$$

Comparing equations (2.62) and (2.63), and using $i^l = e^{il\pi/2}$, we get the scattering amplitude

$$f(\theta) = \frac{1}{2ik}\sum_{l=0}^{\infty}(2l+1)(\eta_l - 1)P_l(\cos\theta)$$

Using the fact that |Incoming wave| = |Outgoing wave|, equation (2.60) gives

$$|\eta_l|^2 = 1 \qquad \text{so that we can take} \qquad \eta_l = e^{2i\delta_l}$$

where δ_l is a real quantity. Thus, we have

$$f(\theta) = \frac{1}{2ik}\sum_{l=0}^{\infty}(2l+1)\ (e^{2i\delta_l} - 1)P_l(\cos\theta)$$

$$= \frac{1}{k} \sum_{l=0}^{\infty} (2l+1)\, e^{i\delta_l}\, \frac{(e^{i\delta_l} - e^{-i\delta_l})}{2i} P_l(\cos\theta)$$

$$= \frac{1}{k} \sum_{l=0}^{\infty} (2l+1)\, e^{i\delta_l} \sin\delta_l\, P_l(\cos\theta)$$

Now, using $\eta_l = e^{2i\delta_l}$ in equation (2.60), we get

$$\psi = \frac{1}{kr} \sum_{l=0}^{\infty} (2l+1) i^l\, \frac{e^{2i\delta_l}\, e^{i(kr - l\pi/2)} - e^{-i(kr - l\pi/2)}}{2i} P_l(\cos\theta)$$

$$= \frac{1}{kr} \sum_{l=0}^{\infty} (2l+1) i^l\, e^{i\delta_l} \frac{e^{i(kr - l\pi/2 + \delta_l)} - e^{-i(kr - l\pi/2 + \delta_l)}}{2i} P_l(\cos\theta)$$

$$= \frac{1}{kr} \sum_{l=0}^{\infty} (2l+1) i^l\, e^{i\delta_l} \sin(kr - l\pi/2 + \delta_l)\, P_l(\cos\theta) \qquad (2.64)$$

Comparison of equations (2.58) and (2.64) shows that the term $\sin(kr - l\pi/2)$ is changed to $e^{i\delta_l}\sin(kr - l\pi/2 + \delta_l)$. Thus, a phase shift is introduced by the scattering potential. Hence, δ_l has physical significance of phase shift in the asymptotic form (large r) of the partial l-wave. Using equation (2.57), total scattering cross section is

$$\sigma = \frac{2\pi}{k^2} \int_0^{\pi} \left| \sum_{l=0}^{\infty} (2l+1)\, e^{i\delta_l} \sin\delta_l\, P_l(\cos\theta) \right|^2 \sin\theta\, d\theta$$

Using the orthonormality relation for Legendre polynomials

$$\int_0^{\pi} P_l(\cos\theta)\, P_m(\cos\theta) \sin\theta\, d\theta = \frac{2}{2l+1}\, \delta_{lm}$$

we get

$$\sigma = \frac{4\pi}{k^2} \sum_{l=0}^{\infty} (2l+1) \sin^2\delta_l$$

For low energy neutrons, only s-wave scattering is predominant and we have

$$\sigma_0 = \frac{4\pi}{k^2} \sin^2\delta_0 \qquad (2.65)$$

Phase shift

For calculation of cross section σ_0, we require the phase shift δ_0 for s-waves ($l = 0$). As we used for deuteron, let us consider a square well potential with depth $-V_0$ and width b.

$$V(r) = \begin{cases} -V_0 & \text{for} \quad 0 \le r < b \\ 0 & \text{for} \quad r > b \end{cases}$$

Using the spherical coordinates, for spherically symmetric potential $V(r)$, the radial part of Schrödinger equation is

$$\frac{1}{r^2}\frac{\mathrm{d}}{\mathrm{d}r}\left(r^2\frac{\mathrm{d}R}{\mathrm{d}r}\right) + \frac{2\mu}{\hbar^2}\Big[E - V(r)\Big]R = 0$$

Using $R = u(r)/r$ here, we get

$$\frac{\mathrm{d}^2u}{\mathrm{d}r^2} + \frac{2\mu(E+V_0)}{\hbar^2}u = 0 \qquad \text{or} \qquad \frac{\mathrm{d}^2u}{\mathrm{d}r^2} + K^2u = 0 \qquad \text{for } r < b \quad (2.66)$$

and

$$\frac{\mathrm{d}^2u}{\mathrm{d}r^2} + \frac{2\mu E}{\hbar^2}u = 0 \qquad \text{or} \qquad \frac{\mathrm{d}^2u}{\mathrm{d}r^2} + k^2u = 0 \qquad \text{for } r > b \quad (2.67)$$

where $K^2 = 2\mu(E + V_0)/\hbar^2$ and $k^2 = 2\mu E/\hbar^2$. Solution of equation (2.66) is

$$u = A\sin Kr + A'\cos Kr \qquad \text{for } r < b$$

where A and A' are constants. For $r \to 0$, we must have $u \to 0$, so that the wave function R is finite at $r = 0$. This can be possible when $A' = 0$ and we have

$$u = A\sin Kr \qquad \text{for } r < b$$

For the region $r > b$, let the solution of equation (2.67) be of the form

$$u = B\sin(kr + \delta_0) \qquad \text{for } r < b$$

where B is a constant and δ_0 is a constant phase shift introduced by the scattering potential. For the conditions of continuity of the wave function u and its derivative at $r = b$, we get

$$A\sin Kb = B\sin(kb + \delta_0) \qquad \text{and} \qquad KA\cos Kb = kB\cos(kb + \delta_0)$$

These relations give

$$K\cot Kb = k\cot(kb + \delta_0) \tag{2.68}$$

For low energy neutrons, it is possible to approximate the quantity $(du/dr)/u$ in the region $r \leq b$ to be equal to some constant. Let it be $-\gamma$. Hence, we have

$$K \cot Kb = -\gamma$$

Thus, equation (2.68) gives

$$k \cot(kb + \delta_0) = -\gamma \qquad \text{or} \qquad \cot(kb + \delta_0) = -\gamma/k$$

Therefore, we have

$$\delta_0 = \cot^{-1}(-\gamma/k) - kb$$

For low energies of neutrons, kb is much smaller than the other terms, and we have

$$\delta_0 = \cot^{-1}(-\gamma/k) \qquad \text{or} \qquad \cot \delta_0 = -\gamma/k$$

Now, we have

$$\operatorname{cosec}^2\delta_0 = 1 + \cot^2 \delta_0 = \frac{k^2 + \gamma^2}{k^2} \qquad \text{therefore,} \qquad \sin^2 \delta_0 = \frac{k^2}{k^2 + \gamma^2}$$

Using this expression in equation (2.65), we have

$$\sigma_0 = \frac{4\pi}{k^2 + \gamma^2}$$

For the deuteron, we have

$$k^2 = \frac{2\mu E}{\hbar^2} \qquad \text{and} \qquad \gamma^2 = \frac{2\mu B}{\hbar^2}$$

where B is the binding energy of the deuteron. Thus, we have

$$\sigma_0 = \frac{2\pi\hbar^2}{\mu(E + B)} \tag{2.69}$$

Equation (2.69) shows that the total scattering cross section σ_0 varies with energy E. The theoretical values of cross sections are found to be much smaller than the experimental ones. For the low energies, the calculated value of cross section is about 5 barns while the experimental value is about 20 barns.

In the calculations, we have used the binding energy and other parameters pertaining to the deuteron which is in the triplet state. In an actual experiment, the spins of neutrons in a beam are oriented at random and the collisions take place where the spins of proton and neutron are, both, parallel as well as anti-parallel. It may be a probable cause for such large discrepancy between theory and experiment.

In 1935, Winger attempted to resolve this discrepancy. He pointed out that though singlet state is unbound in the deuteron, but it should be included in the calculation of scattering cross-sections. In a n-p scattering, the probability of triplet collision ($I = 1$, parallel spins) is three times that of the singlet collision ($I = 0$, anti-parallel spins). These three triplet states and one singlet state have equal probability for their existence. Thus, the total scattering cross section for low energies should be expressed as

$$\sigma = \frac{3}{4}\,\sigma_t + \frac{1}{4}\,\sigma_s$$

where σ_t and σ_s are, respectively, the scattering cross sections for triplet and singlet states and the corresponding weight factors are 3/4 and 1/4. In order to get an agreement between the experimental values and the calculated ones for σ_t, it is possible to estimate σ_s. For the experimental value of 20 barns for total scattering cross section and the calculated value of 5 barns for σ_t, we have

$$20 = \frac{3}{4}\,5 + \frac{1}{4}\,\sigma_s$$

The value of σ_s comes out to be 65 barns. It shows that during the interaction, the singlet state though exists only for one-fourth of the time, the scattering due to it is still quite large.

5.4 Neutron-proton scattering at low energies

In an experiment, a beam of neutrons generated by a nuclear reactor or by some other neutron source is allowed to fall towards a target, containing hydrogenous material having mostly protons. Since, neutrons have no charge, electrons have practically no effect on the incoming neutrons. The target is thin enough so that multiple scattering does not take place. Each of the scattered particle is due to a single interaction. The energy of incident neutrons in the beam is kept much higher than the chemical binding energy of a proton in a molecule, so that the proton appears relatively free to the incoming neutrons.

Theoretical investigation of scattering process cannot be entirely regarded in terms of sharp trajectories of nuclear particles involved in the scattering because of their smallness in size. For the energies of interest, the particles are treated as waves and the scattering process is the mutual interaction between such waves. However, the particle picture is retained in the treatment of the interaction process. Let us consider that a neutron of energy E is moving with velocity v towards a proton (Figure 11). Suppose b be the radius of sphere of influence or the range of the nuclear force. In Figure 11, R, known as the impact parameter, is the perpendicular distance between the direction of neutron and the center of scattering O.

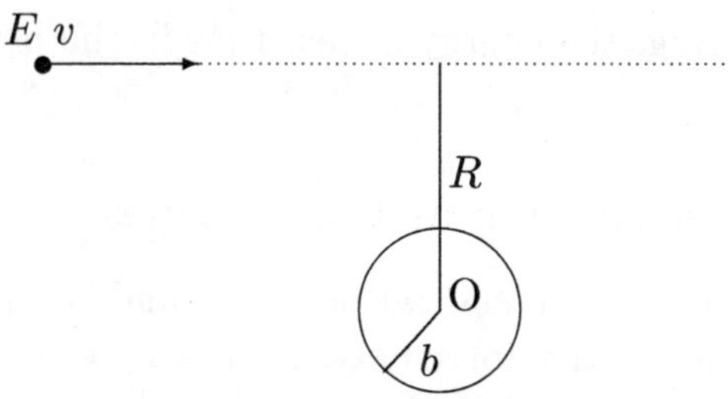

Figure 11: A neutron moving with velocity v approaches the scattering center (proton) O. R is the impact parameter and b the range of nuclear force.

When p is the linear momentum of neutron, classically, the angular momentum of neutron is pR. Quantum mechanically, this angular should be quantized and should be equal to an integer multiple of $\hbar$. Thus, we have

$$pR = l\hbar$$

where l can assume the integer values 0, 1, 2, ... Now, we have

$$l = \frac{2\pi pR}{h}$$

Classically, Figure 11 shows that there will be interaction between a neutron (projectile) and a proton (target), as along as the impact parameter R is smaller than the range b of the nuclear force. Thus, we should have

$$l < \frac{2\pi pb}{h}$$

Using the de Broglie relation $\lambda = h/p$, we have

$$l < \frac{2\pi b}{\lambda}$$

When the energy of incoming neutron is such that the de Broglie waves associated with them have wavelengths $\lambda \geq (2\pi b)$, we have $l < 1$. It therefore shows that essentially the state $l = 0$ will contribute to scattering of those neutrons whose wavelength of the order of the range of nuclear force. For such neutrons, it is said to be s-wave scattering. For

$$\lambda = \frac{h}{\sqrt{2mE}} \qquad \text{we have} \qquad E = \frac{h^2}{2m\lambda^2}$$

Thus. for s-wave scattering, the energy of neutrons in the beam should be less than $h^2/2m\lambda^2$.

5.5 Proton-proton scattering at low energies

The stability of a nucleus having several neutrons and several protons shows that like a strong attractive force between a neutron and a proton, there must be a strong short range attractive force between two protons in a nucleus. Information about nuclear force between two protons is possible only through scattering of a proton by a proton, as no bound state of two protons is available. The proton-proton scattering experiment can be performed with great ease and accuracy. Nowadays, well collimated energetic beams of protons are available from various accelerators. Moreover, protons can be detected easily through the ionization they produce and their energies can be measured accordingly.

However, theoretical analysis of experimental results is rather difficult and complicated, because protons, having electrical charge, experience coulomb scattering in addition to the nuclear scattering. Moreover, we have to account for indistinguishability of two protons, as it is difficult to distinguish between the scattering proton and the scattered proton. Pauli's exclusion principle excludes some states of motion for identical particles, so there will be either singlet or triplet state, but not both the states for a given orbital state (l state) in the p-p scattering. Being fermions, protons obey Fermi-Dirac statistics and they can exist in a state which has anti-symmetric wave function. The two protons, therefore, cannot stay in a state that has symmetric wave function. It shows that a symmetric space wave function (S, D states) can only be associated with an antisymmetric spin wave function and an antisymmetric space wave function (P, F state) can only be associated with a symmetric spin wave function. This shows that for low incident proton energies, the nuclear scattering is s-wave scattering and the interaction is purely singlet-state interaction. There is only one phase shift, unlike two in the n-p scattering (for singlet and triplet).

Being of long range type, the coulomb potential is always effective no matter how distant apart are the colliding particles from each other. As such for the coulomb potential, one calculates differential scattering cross-section as a function of scattering angle. Let us first examine the scattering of protons under coulomb potential alone. Schrödinger equation in the relative coordinates of two protons is

$$\nabla^2\psi + \left(k^2 - \frac{2k\eta}{r}\right)\psi = 0 \qquad \text{with} \quad k = \frac{\mu v}{\hbar} \quad \text{and} \quad \eta = \frac{e^2}{\hbar v}$$

where μ is the reduced mass of the system (it is equal to $m_p/2$, where m_p is mass of proton) and v the relative velocity when the protons are separated by a large distance. As $r \to \infty$, the asymptotic solution of Schrödinger equation is

$$\psi = \text{cons}\left[\mathrm{e}^{-i\eta \ln 2kr\sin^2\theta/2 + ikr\cos\theta} - \frac{\eta}{2kr\sin^2\theta/2}\,\mathrm{e}^{ikr}\mathrm{e}^{-i\eta\ln 2kr\sin^2\theta/2}\mathrm{e}^{2i\eta_0}\right]$$

where

$$\mathrm{e}^{2i\eta_0} = \frac{\Gamma(1+i\eta)}{\Gamma(1-i\eta)} \quad \text{and } \eta_0 \text{ is a real number.}$$

From this solution, we find that the coulomb interaction modifies the phase of both incident and scattered waves. It is therefore not possible to obtain the asymptotic form of the scattered wave as a sum of incident plane wave and a scattered spherical wave of certain angle dependent amplitude. However, the angle dependent term in the modified scattered wave gives the scattering amplitude $f(\theta)$ as

$$f_c(\theta) = \frac{\eta}{2k\sin^2(\theta/2)}\,\mathrm{e}^{-i\eta\ln\sin^2(\theta/2)} \tag{2.70}$$

and the differential scattering cross section in the center-of-mass system is

$$\sigma(\theta) = |f_c(\theta)|^2 = \frac{\eta^2}{4k^2\sin^4(\theta/2)} = \frac{e^4}{4\mu^2v^4}\,\frac{1}{\sin^4(\theta/2)} \tag{2.71}$$

This is Rutherford formula for scattering of a proton ($Z = 1$) by a proton. Initially Rutherford derived this formula on the basis of classical dynamics. In an actual experiment on proton scattering, the protons scattered at an angle θ and at an angle $(\pi - \theta)$ cannot be distinguished from the recoiled protons, in particular when the spin is neglected. The scattering amplitude for a proton scattered at $(\pi - \theta)$ is

$$f_c(\pi - \theta) = \frac{\eta}{2k\cos^2(\theta/2)}\,\mathrm{e}^{-i\eta\ln\cos^2(\theta/2)} \tag{2.72}$$

and the differential scattering cross section in the center-of-mass system is

$$\sigma(\pi - \theta) = |f_c(\pi - \theta)|^2 = \frac{e^4}{4\mu^2 v^4} \frac{1}{\cos^4(\theta/2)} \tag{2.73}$$

According to classical theory, the differential scattering cross section for indistinguishable particles is the sum of $\sigma(\theta)$ and $\sigma(\pi-\theta)$. Therefore, the differential scattering cross section for coulomb scattering of proton at an angle θ in the center-of-mass system is

$$\sigma_c(\theta) = \sigma(\theta) + \sigma(\pi - \theta) = \frac{e^4}{4\mu^2 v^4}\left[\frac{1}{\sin^4(\theta/2)} + \frac{1}{\cos^4(\theta/2)}\right] \tag{2.74}$$

But in quantum mechanics, the indistinguishability of protons puts another point for consideration. The wave functions for the protons should be properly symmetrised. Instead of taking sum of the squares of the scattering amplitudes as performed in equations (2.71) and (2.73), we should take a linear combination of the scattering amplitudes $\sigma(\theta)$ and $\sigma(\pi - \theta)$. The wave functions are combined linearly such that the total wave function is antisymmetric, as the protons are fermions. If the space part of the combination is symmetric then the spin part is antisymmetric; if the space part is antisymmetric then the spin part is symmetric. For the triplet state ${}^3S_1$, the spin wave function is symmetric and therefore the space wave function should be antisymmetric. The combination $f(\theta) - f(\pi - \theta)$ is antisymmetric. For the singlet state ${}^1S_0$, the spin wave function is antisymmetric and therefore the space wave function should be symmetric. The combination $f(\theta) + f(\pi - \theta)$ is symmetric. Further, the scattering at θ and $(\pi - \theta)$ is three times more frequent in the triplet state than in the singlet state. Thus, the differential scattering cross section is

$$\sigma(\theta) = \frac{1}{4}\left[f(\theta) + f(\pi - \theta)\right]^2 + \frac{3}{4}\left[f(\theta) - f(\pi - \theta)\right]^2$$

$$= |f(\theta)|^2 + |f(\pi - \theta)|^2 - |f(\theta)^* f(\pi - \theta)| \tag{2.75}$$

Using equations (2.70) and (2.72) in (2.75), we get the differential scattering cross section as

$$\sigma(\theta) = \frac{e^4}{4\mu^2 v^2}\left[\frac{1}{\sin^4(\theta/2)} + \frac{1}{\cos^4(\theta/2)} - \frac{\cos\{\eta \ln \tan^2(\theta/2)\}}{\sin^2(\theta/2)\,\cos^2(\theta/2)}\right] \tag{2.76}$$

This equation (2.76) is the Mott scattering formula for protons under coulomb force. Here, the third term arises from the quantum treatment of coulomb scattering for indistinguishable charged particles (protons). For the proton energies of interest (above 1 MeV), η is small so that the cosine term in the numerator of the third term is of the order of unity unless $\theta = 0$ or π. Thus, when θ is neither zero or π and for the energies larger than 1 MeV, we have

$$\sigma(\theta) = \frac{e^4}{4\mu^2 v^2}\Big[\frac{1}{\sin^4(\theta/2)} + \frac{1}{\cos^4(\theta/2)} - \frac{1}{\sin^2(\theta/2)\,\cos^2(\theta/2)}\Big] \quad (2.77)$$

When the distance between two protons is of the order of nuclear force, nuclear scattering becomes effective. In actual computation of *p-p* scattering, it is reasonable to assume that the nuclear potential for *p-p* scattering is the same as that for the *n-p* scattering. Due to the Pauli's exclusion principle, the triplet state is excluded for the identical particles and therefore the lowest state, $l = 0$ is a singlet state. Thus, at low energies, we have the contribution of *s*-wave scattering only, as was in the case of *n-p* scattering. The differential cross section when computed in the center-of-mass system, under the joint effect of coulomb and nuclear potentials, is

$$\sigma(\theta) = \frac{e^4}{4\mu^2 v^2}\Big[\frac{1}{\sin^4(\theta/2)} + \frac{1}{\cos^4(\theta/2)} - \frac{1}{\sin^2(\theta/2)\,\cos^2(\theta/2)}\Big] + \frac{\sin^2\delta_0}{k^2}$$

$$-\frac{e^2}{2\mu v^2}\frac{\sin\delta_0}{k}\Big\{\frac{\cos(\eta \ln \sin^2(\theta/2) + \delta_0)}{\sin^2(\theta/2)} + \frac{\cos(\eta \ln \cos^2(\theta/2) + \delta_0)}{\cos^2(\theta/2)}\Big\} \quad (2.78)$$

Since η is small for energies of the order of 1 MeV and becomes still smaller at higher energies, under this consideration equation (2.78) reduces to

$$\sigma(\theta) = \frac{e^4}{4\mu^2 v^2}\Big[\frac{1}{\sin^4(\theta/2)} + \frac{1}{\cos^4(\theta/2)} - \frac{1}{\sin^2(\theta/2)\,\cos^2(\theta/2)}\Big]$$

$$+\frac{\sin^2\delta_0}{k^2} - \frac{e^2}{2\mu v^2}\frac{\sin\delta_0\cos\delta_0}{k\sin^2(\theta/2)\cos^2(\theta/2)} \quad (2.79)$$

In this equation (2.79), the first term is the Mott scattering (pure coulomb scattering), the second term is pure nuclear potential scattering (*s*-wave scattering) and the last term corresponds to the interference between the coulomb and nuclear scattering, which depends linearly on the phase shift δ_0.

5.6 Neutron-neutron scattering

Similar to the case of *p-p* scattering, there is no bound system where only two neutrons exist. Thus, it is not possible to study the *n-n* interaction directly. By bombarding D_2O and H_2O with neutron of 300 MeV, the values of differential scattering cross section obtained are:

$$\sigma_{nn}(\theta) = 3.7 \text{ mb/str} \qquad \sigma_{nn} = 22 \text{ mb}$$

$$\sigma_{pp}(\theta) = 3.6 \text{ mb/str} \qquad \sigma_{pp} = 23 \text{ mb}$$

Since total cross section for n-D_2O scattering and for n-H_2O scattering may be expressed as

$$\sigma_{n\text{-}\mathrm{D_2O}} = \sigma_{nn} + \sigma_{np} + I$$

$$\sigma_{n\text{-}\mathrm{H_2O}} = \sigma_{pp} + \sigma_{np} + I$$

where I denotes the interaction. Such investigations show the equality of n-n and p-p interactions. The idea about the equality of n-n and p-p interactions can also be obtained from a comparison of binding energies of lightest pair of mirror nuclei. However, in the case of p-p interactions, one also has coulomb part which when removed from the total interaction, we get an interaction with is equal to the n-n interaction.

Note: This supports the hypothesis that the nuclear forces between two nucleons do not depend on charge, *i.e.*, they are charge independent. Thus, the nuclear force between a proton and a neutron (n-p) is the same as between two protons (p-p) as well as between two neutrons (n-n). That is, n-n = n-p = p-p. The equality of nuclear force between two protons and between two neutrons shows that the nuclear forces are charge symmetric (n-n = p-p).

The charge independence of nuclear forces has given an entirely new formalism. The two particles, proton and neutron, are considered to be two charge-states of a single particle, normally called as the nucleon. Heisenberg was the first to associate the idea of isotopic spin or iso-spin with this charge independence of nuclear forces.

The idea is similar to ordinary spin. Just as a proton can have two spin states ($m_s = \pm 1/2$), a nucleon has two states of iso-spin ($T_3 = \pm 1/2$), $T_3 = 1/2$ state representing a proton and $T_3 = -1/2$ state representing a neutron. Thus, a p-p system with $T_3 = 1$ is necessarily an iso-triplet with $T = 1$ while an n-p system with $T_3 = 0$ can be an iso-singlet with $T = 0$ or an iso-triplet with $T = 1$.

6. Problems and questions

1. For deuteron, spherically symmetric potential is expressed as

$$V(r) = \begin{cases} -V_0 & \text{for} \quad 0 \leq r < b \\ 0 & \text{for} \quad r > b \end{cases}$$

   where b is the range of the nuclear force. Discuss the fact that the deuteron is loosely bound.

2. Discuss that a deuteron has a ground S state and cannot have any excited S state.

3. Explain that a deuteron does not have any excited P state.

4. Describe the laboratory and center-of-mass coordinate systems. Obtain relation between the elastic scattering cross sections in the two coordinate systems.

5. For an elastic scattering, obtain the relation between the differential scattering cross section and amplitude of the scattered wave.

6. Show that for s-wave scattering of a beam of neutrons by a proton, the energy of neutrons should be less than $h^2/2m\lambda^2$, where λ is the de Broglie wavelength associated with a neutron.

7. Write notes on the following

   (i) Charge symmetry of nuclear forces
   (ii) Yukawa potential
   (iii) Exchange force
   (iv) Laboratory and center-of-mass coordinate systems
   (v) Neutron-neutron scattering
   (vi) Proton-proton scattering

III. Nuclear Models

A nucleus consists of a number of nucleons (protons and neutrons) which interact with one another due to strong interactions, called the nuclear forces. The nature of these interactions is not known exactly. Thus, one cannot write Schrödinger equation for a nucleus and solve it. Under such situation, to understand the properties of a nucleus, one has to rely on some simple model for the nucleus. Tremendous amount of experimental data has been used to find out similarities and patterns which could help in the development of some model for the nucleus. Niel Bohr gave the first model, called the liquid-drop model, proposing that a nucleus behaves like a drop of liquid. In order to explain particular characteristics shown by some nuclei having specific number of nucleons, shell model of nucleus has been proposed. Besides, these two models, a collective model for a nucleus has been proposed. In this chapter, we shall discuss about some models proposed for a nucleus.

1. Liquid drop model

Niel Bohr noticed that as different amount of a given liquid has the same density under the same conditions, nuclei of various atoms also have the same density. Therefore, in 1935 he proposed the liquid-drop model for a nucleus. The idea for this model is based on the following similarities.

1. The fact that the density of nuclei of all atoms is almost the same shows the resemblance to a liquid drop. For a given liquid, the density does not depend on the amount of the liquid.

2. The process of disintegration of nuclei through emission of particles is can be taken similar to the evaporation of molecules from the surface of a liquid drop.

3. For a large number of stable nuclei, the binding energy per nucleon is almost the same. This property is similar to the latent heat of evaporation of liquid.

4. The formation of a compound nucleus through absorption of bombarding particles is very similar to the condensation of drops.

5. The energy of nuclei corresponds to the internal thermal motions in a drop of liquid.

According to this model, a stable nucleus is considered spherical in shape. The surface tension forces acting on the surface of a liquid find a similarity as the potential barrier effects on the surface of a nucleus. Like the molecules in a liquid drop, the nucleons in the nucleus interact only with their immediate neighbours. Independence of the binding energy per nucleon for a large number of stable nuclei resembles the independence of the heat of vaporization of a liquid drop on its size. The molecules evaporate from the liquid drop on increase of the temperature of the liquid due to their increased thermal energy. Similarly, when an energy is given to a nucleus by bombarding it with nuclear particles, a compound nucleus is formed which then emits nuclear radiations.

The phenomenon of nuclear fission is analogous to the splitting of a liquid drop into two parts of nearly equal size when it is made to oscillate. Thus, the phenomenon of nuclear fission can be explain with the help of liquid drop model of a nucleus.

The liquid-drop model has been used for calculation of binding energy of a nucleus. For details, please see chapter I.

Exercise 1. Calculate the excitation energy of U^{236} nucleus, when U^{235} absorbs a thermal neutron of energy $E = 0.025$ MeV. Given, masses of U^{235}, U^{236} and neutron are, respectively, 235.0439 amu, 236.0457 amu and 1.0087 amu.

Solution: Mass of U^{235} and neutron = 235.0439 + 1.0087 = 236.0526 amu

For the formation of compound nucleus

$$U^{235} + n^1 \rightarrow U^{236}$$

Mass defect Δm = 236.0526 - 236.0457 = 0.0069 amu

Equivalent energy = 0.0069 × 931 = 6.4239 MeV

Average binding energy per nucleon is

$$\frac{6.4239}{236} = 0.0272 \text{ MeV/nucleon}$$

Excitation energy = Average binding energy per nucleon + Kinetic energy of captured neutron = 0.0272 + 0.025 = 0.0522 MeV.

2. Quantum numbers for individual nucleons

Before going to the shell model of a nucleus, let us first discuss about various quantum numbers for individual nucleons. Various states of a system are expressed through its wave-functions. The solutions (wave-functions) of the wave equation for a system are characterized with the help of some numbers, known as the quantum numbers. Because of many apparent similarities between a atomic system and a nuclear system, it seems logical to use the same nomenclature for the quantum numbers for a nucleus as for an atom. In an atom, we have a system of electrons whereas in a nucleus we have nucleons (protons and neutrons). In a nucleus, protons and neutrons have similar kind nuclear forces.

2.1 Principle quantum number n

For each bound particle, there is a quantum number, called the principle quantum number, denoted by n. It can assume positive integer values, 1,2,3, ... In a strong central field, the first order term for total energy of the state is inversely proportional to n^2. It is expressed as

$$E_n = -\frac{\text{constant}}{n^2}$$

2.2 Orbital quantum number l

As in the atomic physics, for a given value of n, the orbital quantum number l can assume positive integer values, ranging from 0 to $(n - 1)$ with the steps of unit. That is, l can assume the values, 0, 1, 2, ..., $(n - 1)$. The corresponding orbital angular momentum obtained quantum mechanically is $\hbar\sqrt{l(l+1)}$. The values $l = 0, 1, 2, 3, 4, \ldots$ are designated, respectively, by the letters, s, p, d, f, ...

As in the atomic physics, for a given value of l, the orbital magnetic quantum number m_l denotes the projection of $l^* = \sqrt{l(l+1)}$ on the axis of quantization. This axis of quantization is generally taken as the $z-$axis along which external fields also are applied. For a given l, m_l can assume $(2l + 1)$ values, ranging from $-l$ to $+l$ with the steps of unity. That is, m_l can assume the values, $-l$, $(-l + 1)$, $(-l + 2)$, ..., $(l - 1)$, l.

2.3 Spin quantum number s

The intrinsic angular momentum associated with a proton as well as a neutron is known as the spin and is represented by a quantum number s. Both proton as well as neutron are fermions and the value of the spin quantum number s is 1/2 in each case.

As in the atomic physics, for a given value of s, the spin magnetic quantum number m_s denotes the projection of $s^* = \sqrt{s(s+1)}$ on the axis of quantization. This axis of quantization is generally taken as the $z-$axis along which external fields also are applied. For a given s, m_l can assume $(2s + 1)$ values, ranging from $-s$ to $+s$ with the steps of unity. Since $s = 1/2$, m_l can assume the values, $-1/2$ and $1/2$.

2.4 Total angular quantum number j

From atomic physics, we are aware that the total angular momentum is the vector sum of its orbital angular momentum and spin angular momentum. It is represented by a quantum number j, so that

$$\vec{j} = \vec{l} + \vec{s}$$

As in the atomic physics, for a given value of j, the magnetic quantum number m_j denotes the projection of $j^* = \sqrt{j(j+1)}$ on the axis of quantization. This axis of quantization is generally taken as the $z-$axis along which external fields also are applied. For a given j, m_j can assume $(2j + 1)$ values, ranging from $-j$ to $+j$ with the steps of unity. That is, m_j can assume the values, $-j$, $(-j+1)$, $(-j+2)$, $\ldots$, $(j-1)$, j.

2.5 Isospin quantum number T

Though a proton has positive charge and a neutron is neutral, but inside a nucleus, they are very similar in all respects. For example, the nuclear forces between them are charge independent. That is a nuclear force between a proton and a neutron is the same as between two protons and between two neutrons. On the basis of that, Heisenberg suggested proton and neutron are two forms of a common particle, called the nucleon. The particle nucleon was assigned a spin, called iso-spin with value 1/2 and denoted by T. This spin has nothing to do with the intrinsic spin of proton or neutron. The projection T_z of iso-spin on the axis of quantization can assume $(2T + 1)$ values. For individual nucleon, the iso-spin is T is 1/2. Thus, T_z can assume two values, $-1/2$ and $1/2$. These values correspond, respectively, to neutron and proton. That, for a proton, $T_z = 1/2$ whereas for a neutron, $T_z = -1/2$.

2.6 Parity

Parity is the property of wave function describing a quantum mechanical system. Suppose on reflection of coordinates about the origin, we may get

$$\psi(-\vec{r}) = \psi(\vec{r}) \qquad \text{even or positive parity}$$

or

$$\psi(-\vec{r}) = -\psi(\vec{r}) \qquad \text{odd or negative parity}$$

This way odd and even parity of the wave function of a system is defined. The reflection about the origin in spherical coordinates is equivalent to replacing θ by $\pi - \theta$ and ϕ by $\pi + \phi$. The function $e^{im\phi}$ becomes $e^{im\phi}\, e^{im\pi} = (-1)^m e^{im\phi}$. For the function $P_{lm}(\theta)$, we have $P_{lm}(\pi - \theta) = (-1)^{l-m} P_{lm}(\theta)$. Hence, the reflection about the origin is equivalent to multiply the wave function by $(-1)^{l-m}(-1)^m = (-1)^l$. It can be concluded that a wave function with even l has an even parity and that with odd l has an odd parity.

It is straightforward that a product of wave functions with l_1, l_2, l_3, ... has even (odd) parity when the sum $l_1 + l_2 + l_3 + \ldots$ is even (odd).

3. Nuclear shell model

Under this concept, a nucleus is considered to be consists of various shells.

3.1 Fermi gas model

In this model, the nucleons in a nucleus are assumed to move in a central potential of a square well type with a depth of about 50 MeV. Square well shape for the potential is the simplest approach to deal with the situation. Further, it is assume that there is no interaction between the individual nucleons. Since the nucleons are Fermi particles, a nucleus is therefore assumed to consist of a Fermi gas of neutrons and protons independently obeying the Pauli's exclusion principle. Thus, the various states are filled from the lowest value. The highest occupied state for each type of particle is known as the Fermi level with energy E_F.

For a nucleon with a momentum between p and $p + \mathrm{d}p$ and confined in a volume V, the volume in the phase space is expressed as $4\pi p^2\, \mathrm{d}p\, V$. The number of nucleon states $\mathrm{d}n$ available to nucleons is

$$\mathrm{d}n = \frac{4\pi p^2 \mathrm{d}p V}{h^3} \tag{3.1}$$

Assuming the coulomb interactions between the protons to be negligible as compared to the nuclear forces and integrating equation (3.1) up to the Fermi level momentum $p_F(=\sqrt{2mE_F})$, where m is the mass of a nucleon, we get

$$n = \frac{4\pi V}{h^3}\int_0^{p_F} p^2\,\mathrm{d}p = \frac{4\pi V p_F^3}{3h^3} \tag{3.2}$$

Since each state can be occupied by 4 nucleons, protons with spin-up and spin-down, and neutrons with spin-up and spin-down, we take $n = A/4$, where A is the number of nucleons. Then equation (3.2) gives

$$p_F^3 = \frac{3Ah^3}{16\pi V} \tag{3.3}$$

Each state can accommodate two neutrons and two protons (spin up and spin down) and the nuclear volume V can be expressed as $V = 4\pi r_0^3/3$. Hence, we have

$$p_F^3 = \frac{9h^3}{64\pi^2 r_0^3} \quad \text{and} \quad E_F = \frac{p_F^2}{2m} = \frac{(9\pi)^{2/3}h^2}{32\pi^2 m r_0^2} \tag{3.4}$$

Using $h = 6.626 \times 10^{-34}$ J s, $m = 1.67 \times 10^{-27}$ kg, $r_0 = 1.2 \times 10^{-15}$ m, we get

$$E_F = \frac{(9\pi)^{2/3} \times (6.626 \times 10^{-34})^2}{32\pi^2 \times 1.67 \times 10^{-27} \times (1.2 \times 10^{-15})^2}$$

$$= 5.365 \times 10^{-12} \text{ J} = 33.49 \text{ MeV}$$

After adding the binding energy of nucleon (= 8 MeV), we get the depth of the potential well as 41 MeV. Due to coulomb repulsive potential, the well depth for protons will be less than that for the neutrons. The asymmetry energy observed in the liquid drop model is explained in this model. This model can be used for calculation of level density of a nucleus but not the exact level energies.

3.2 Single particle shell model

The nuclear shell model is one of the most important and useful models for a nucleus. The idea for this model was derived on the basis of the results of empirical correlations of certain nuclear data. The experimental facts show that those nuclei are more stable for which either the number of protons Z or the numbers of neutrons $N = A - Z$ is equal to one the numbers 2, 8, 20, 50, 82, 126. These numbers are commonly known as the 'magic numbers'.[1] The magic numbers of neutrons and protons have been interpreted as forming closed shells of neutrons or protons in analogy with filling of electron shells in atoms. Further, the neutron and proton shells appear to be independent of each other. This model successfully explains the magic numbers as well as the other nuclear properties such as spin, magnetic moment and energy levels. Some of the evidences which support of this model are as the following:

1. It was found that nuclei having one of the magic numbers of nucleons have large abundance in nature.

2. He^4 ($Z = 2$, $N = 2$) and O^{16} ($Z = 8$, $N = 8$) are particularly stable, as can be seen from the binding energy curve. Here, the numbers 2 and 8 indicate large stability.

3. The most stable and the largest group of isotones (N constant) occurs at $N = 82$. The next groups are at $N = 50$ and $N = 20$. The neutron numbers 20, 50 and 82 therefore indicates large stability.

4. $_{50}Sn$ has ten stable isotopes, while $_{20}Ca$ has six stable isotopes. This indicates that the elements with $Z = 50$ and $Z = 20$ are more stable as compared to others.

5. No more than 5 isotones occurs in nature for any N, except the $N = 50$, where there are six and the $N = 82$, where there are seven. Neutron numbers of 82 and 50, therefore indicate a specific type of stability.

6. The three main radioactive series (uranium, actinium and thorium series) all decay to lead. The most stable isotope of lead with $Z = 82$ and $N = 126$.

7. Nuclei with the $Z = 50$ (Sn) and $Z = 82$ (Pb) have much smaller capture cross sections than their neighbours. Thus, $Z = 50$ and 82 have some specific characteristic.

8. The nuclei with two magic numbers (Z and N both magic numbers), for example, $_2He^4$, $_8O^{16}$, $_{20}Ca^{40}$ and $_{82}Pb^{208}$ are tightly bound.

9. The electric quadrupole moments of nuclei show sharp minimal at the closed shell numbers, indicating that such nuclei are nearly spherical.

[1] This name 'magic numbers' is not a strictly scientific one. It serves the purpose of identifying the numbers simply and conveniently.

Radioactive decay of uranium, actinium and thorium

Three series showing the radioactive decay of uranium, actinium and thorium give the final product as lead with $Z = 82$ and $N = 126$. For computation of shell structure of electrons in an atom, the electrons are considered to move in the electrostatic field of nucleus of the atom. For computation of shell structure of nucleus, each nucleon in the nucleus is supposed to move independently of others in the field representing the average effect of interactions of all the other nucleons in the nucleus. Each nucleon with its spin 1/2 behaves like an independent particle, subjected to the additional condition, imposed by Pauli's exclusion principle. In this 'independent particle model', it is assumed that there is a strong spin-orbit coupling for each individual nucleon.

We now consider a particle of mass m executing simple harmonic motion isotropically in a three dimensional space. The particle is held by the restoring force kr, where the constant k is known as the force constant. The particle will oscillate about the equilibrium position with frequency $\nu = (1/2\pi)\ \sqrt{(k/m)}$ or angular velocity $\omega = \sqrt{(k/m)}$. Now, the potential energy $V(r)$ of the oscillating particle is

$$V(r) = \int_0^r kr\ \mathrm{d}r = \frac{1}{2}\ kr^2 = \frac{1}{2}\ m\omega^2 r^2$$

The Schrödinger equation for the system is

$$\left(-\frac{\hbar^2}{2m}\ \nabla^2 + \frac{1}{2}\ m\omega^2 r^2\right)\psi(\vec{r}) = E\psi(\vec{r})$$

The energy eigen values obtained are

$$E_{nl} = \left(N + \frac{3}{2}\right)\hbar\omega$$

where $N = 2(n-1) + l$ represents the total number of oscillator quanta excited and assume the values $N = 0, 1, 2, 3, \ldots$. They respectively correspond to the ground, first excited, second excited, ..., state. The quantum number n can assume the values $n = 1, 2, 3, \ldots$ and the quantum number l can assume the values $l = 0, 1, 2, \ldots, n$. When the angular dependence of wave functions was examined, it was found that for each value of N, there is a degenerate group of levels with different l-values, such that $l \leq N$. Moreover, l can assume only even (odd) values corresponds to even (odd) values of N. Each value of N represent a shell and for a given value of N, the state can accommodate $(N + 1)(N + 2)$ neutrons or the same number of protons. For a closed shell nucleus where all the shells up to a given N_0 (maximum number of excited quanta) are filled, the total number of nucleons of a particular type (protons or neutrons) will be[2]

$$\sum_{0}^{N_0}(N+1)(N+2) = \frac{1}{3}\,(N_0+1)(N_0+2)(N_0+3)$$

Thus, the sequence of the single particle levels for the harmonic oscillator will consists of bands of degenerate levels, each band separated by the energy $\hbar\omega$ from the next one. The ground state energy is $3\hbar\omega/2$. The degeneracy corresponding to each value of l is $2(2l+1)$, as before. The single particle level scheme predicted by the harmonic oscillator is given in Table 1. In Table 1, we have also given the values of $2(2l+1)$. Against each N, we have also given the values of $(N+1)(N+2)$. It can be noticed that for a given N, the value of $(N+1)(N+2)$ is the sum of all values of $2(2l+1)$.

Since the smaller n should be lower (small energy), we find that the lower l is higher (large energy). Now, owing to the spin-orbit interaction, each nucleon energy level with a given value of l for its orbital quantum number splits into two sub-levels with the total angular quantum number j having the value $l+1/2$ and $l-1/2$, except when $l=0$, in which case j takes only one value, viz., $j = 1/2$. Since the $j = l+1/2$ sub-level has a lower energy than the $j = l-1/2$ sub-level, the former represents a state with a more tightly bound nucleon.

The lowest state $1s_{1/2}$ holds two nucleons. Hence, there is a closed shell at 2. In the next state the interaction is not strong enough to separate the $1p_{3/2}$, and $1p_{1/2}$ levels by an appreciable amount. Consequently, the next closed shell occurs, when all the $6p$ states are filled. Thus, we get the magic number at $2+6=8$. The state $1d_{5/2}$ is definitely lower than $1d_{3/2}$, but it still forms part of the same shell. The states $1d_{5/2}$, $1d_{3/2}$ and $2s_{1/2}$ are sufficiently close to make a single shell. The next shell therefore, closes at $8+12=20$. For the state $1f_{7/2}$ is high enough to the lower state below all the other $n = 3$ states, but not enough,

---

[2]We have

$$\sum_{0}^{N_0}(N+1)(N+2) = \sum_{0}^{N_0}(N^2+3N+2) = \sum_{0}^{N_0} N^2 + 3\sum_{0}^{N_0} N_0 + 2(N_0+1)$$

$$= \frac{1}{6}\,N_0(N_0+1)(2N_0+1) + \frac{3}{2}\,N_0(N_0+1) + 2(N_0+1)$$

$$= \frac{1}{3}\,(N_0+1)(N_0+2)(N_0+3)$$

and therefore it joins the group from $n = 2$. Thus, this shell closes at $20 + 8 = 28$.

Table 1. Level scheme predicted by the harmonic oscillator

N	E_N	l	n	nl	levels	$2(2l+1)$	$(N+1)$ $\times(N+2)$
0	$3\hbar\omega/2$	0	1	$1s$	$1s_{1/2}$	2	2
1	$5\hbar\omega/2$	1	1	$1p$	$1p_{1/2}$, $1p_{3/2}$	6	6
2	$7\hbar\omega/2$	2	1	$1d$	$1d_{3/2}$, $1d_{5/2}$	10	12
		0	2	$2s$	$2s_{1/2}$	2	
3	$9\hbar\omega/2$	3	1	$1f$	$1f_{5/2}$, $1f_{7/2}$	14	20
		1	2	$2p$	$2p_{1/2}$, $2p_{3/2}$	6	
4	$11\hbar\omega/2$	4	1	$1g$	$1g_{7/2}$, $1g_{9/2}$	18	30
		2	2	$2d$	$2d_{3/2}$, $2d_{5/2}$	10	
		0	3	$3s$	$3s_{1/2}$	2	
5	$13\hbar\omega/2$	5	1	$1h$	$1h_{9/2}$, $1h_{11/2}$	22	42
		3	2	$2f$	$2f_{5/2}$, $2f_{7/2}$	14	
		1	3	$3p$	$3p_{1/2}$, $3p_{3/2}$	6	
6	$15\hbar\omega/2$	6	1	$1i$	$1i_{11/2}$, $1i_{13/2}$	26	56
		4	2	$2g$	$2g_{7/2}$, $2g_{9/2}$	18	
		2	3	$3d$	$3d_{3/2}$, $3d_{5/2}$	10	
		0	4	$4s$	$4s_{1/2}$	2	
7	$17\hbar\omega/2$	7	1	$1j$	$1j_{13/2}$, $1j_{15/2}$	30	72
		5	2	$2h$	$2h_{9/2}$, $2h_{11/2}$	22	
		3	3	$3f$	$3f_{5/2}$, $3f_{7/2}$	14	
		1	4	$4p$	$4p_{1/2}$, $4p_{3/2}$	6	

The next shell compromises $1f_{5/2}$, $2p_{3/2}$, $2p_{1/2}$ and $1g_{9/2}$. The last one coming down because of spin-orbit coupling from the next higher group of levels. As it contains 22 sub-levels, this shell closes at $28+22 = 50$. Similarly the next shell is made up of the 32 sub-levels of $1g_{7/2}$, $2d_{5/2}$, $2d_{3/2}$, $3s_{1/2}$ and $1h_{11/2}$, and this shell closes at $50 + 32 = 82$ and the one after that with 44 sub-shells of $1h_{9/2}$, $2f_{7/2}$, $2f_{5/2}$, $3p_{3/2}$, $3p_{1/2}$ and $1i_{13/2}$ closes at $82 + 44 = 126$.

In this way we see that the shell closures takes place at particle numbers 2, 8, 20, 28, 50, 82, 126 exactly as required by experiments. These number are the magic numbers.

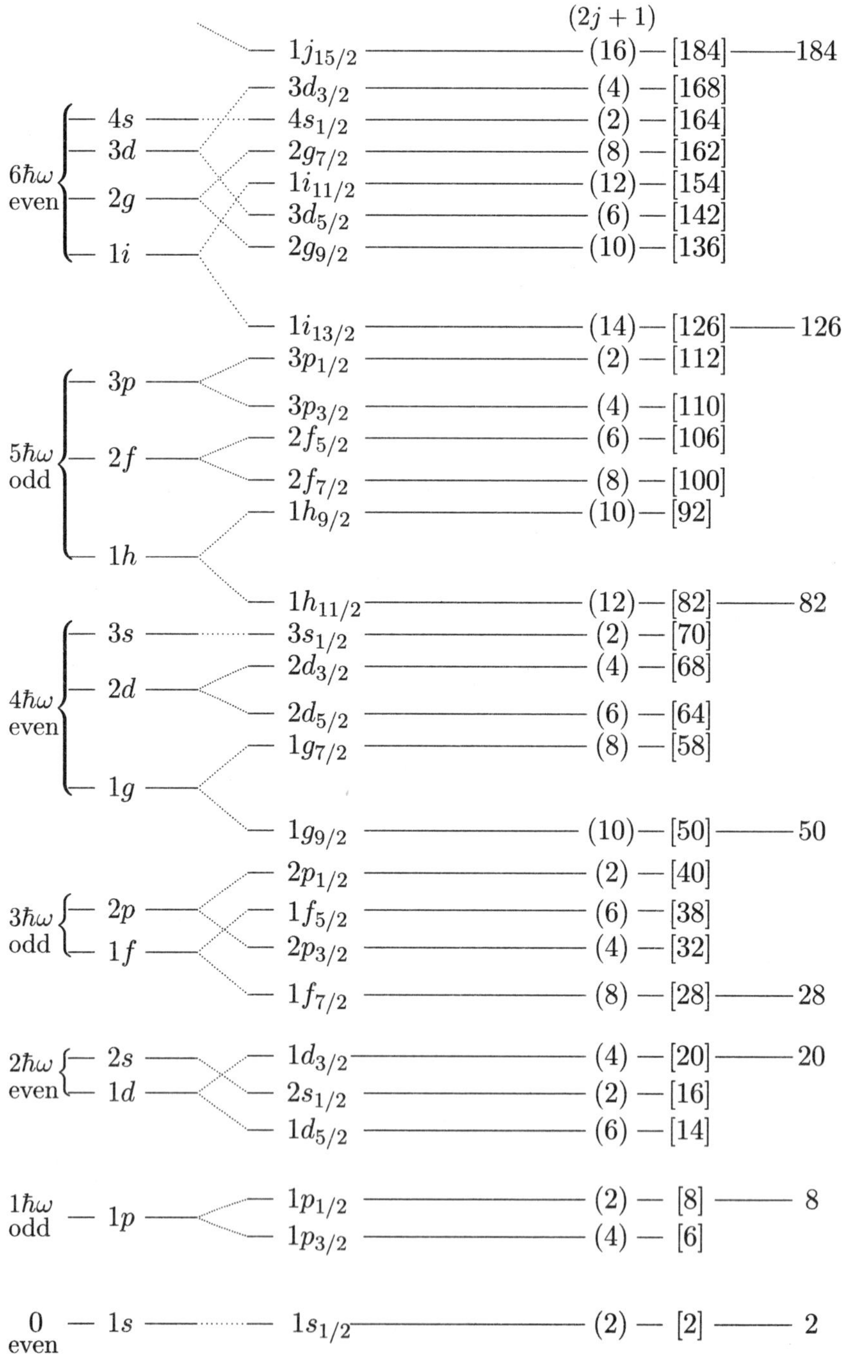

Figure 1: The energies are given relative to the ground state. The values in the pairs of brackets are $(2j+1)$. The values in the pairs of square brackets are the cumulative values. The magic numbers are given on the far right.

Predictions of shell model

1. **Stability of closed shell nuclei:** Shell model of a nucleus successfully produces all magic numbers, supporting the idea for the stability of closed shell nuclei.

2. **Spin of nuclear ground state:** The shell model has been very successful in predicting the ground state spin in case of a large number of nuclei. According to the shell model, the levels for the neutrons are filled independent of the levels for the protons.

3. **Magnetic moment of a nucleus:** In an odd nucleus, the total angular momentum I of the nucleus is equal to the angular momentum j of the last unpaired nucleon. Thus we see that magnetic moment of the nucleus is produced by the odd nucleon only.

4. **Electric quadrupole moment of a nucleus:** The shell model makes predictions about the electric quadrupole moment of a odd A, odd Z nuclide, on the assumption that this is due to the unpaired proton.

5. **Nuclear isomerism:** The shell model for a nucleus successfully explained the phenomenon of nuclear isomerism.

6. **Beta decay:** One more application of shell model is the understanding of beta decay. Theory of β-decay shows that the life time can be understood in terms of relative parity and angular momenta of the states involved. The nuclear shell model is capable to predict the spins and parity of these states, of unstable nuclei which decay through the β-emission.

7. **Stripping reactions:** The nuclear shell model can be used to understand the stripping reactions.

4. Collective nuclear model

We have seen that the liquid-drop model considers the interactions between the nucleons inside a nucleus and thus it accounts for the collective behaviour of nucleons. In this model, the nucleons are considered to interact with the nearest neighbours, as the nuclear forces are of short range. This model, for example, could explain the phenomenon of fission successfully. But, it cannot explain, for example, the phenomenon of single neutron emission. On the other side, in the shell model, we do not consider any interaction between the nucleons in a nucleus. In the nuclear shell model, the nucleons are treated individually. In this model we can explain specific behaviour of nuclei having magic numbers for nucleons. This model can also explain the the phenomenon of single neutron emission. But, this model cannot explain the phenomenon of fission. It shows that each molecule has its usefulness and is applicable for some nuclear processes and not applicable for other processes.

Scientists made attempt to combine these two models together so that a model having features of both could be achieved. This model is known as the 'Collective Model'.

In this model, the particles within the nucleus exert a centrifugal pressure on the surface of the nucleus and as a result of that, the nucleus may be deformed into a permanently non-spherical shape. The surface may undergo oscillations (liquid-drop aspect). The particles within the nucleus then move in a non-spherical potential like that assumed in order to account for the quadrupole moments. This way, the nuclear distortion reacts on the particles and modifies somewhat the independent-particle aspect. Now, the nucleus is considered as having the shell structure capable of performing oscillations in the shape and size. Consequently, the collective model can be made to describe such drop-like properties as nuclear fission and on the same time it preserves the shell model characteristics, in fact, improving on the earlier shell model.

The scheme of nuclear energy levels which is consequence of the collective motion of nucleons in the core and the motion of loosely bound surface nucleons depends upon the strength of the coupling between them. For the strong coupling, the energy states resemble with those for the linear molecules. Corresponding to the rotation, vibration and electronic energy states of a molecule, we have rotational, vibrational and nucleonic energy states of a nucleus. The rotation and vibration arise from the motion of the nuclear core whereas the nucleonic states arise due to the motion of the loosely bound nucleons. The total energy may be expressed as

$$E_T = E_{rot} + E_{vib} + E_{nuc}$$

The simplest type of collective motion which has been identified experimentally is related with the rotations of the deformed nucleus. When the rotational collective motion is sufficiently slow, it will not affect the internal structure of the nucleus. Following the classical mechanics, the rotational energy is proportional to the square of the angular velocity ω, so that

$$E_{rot} = \frac{1}{2}\, I\omega^2$$

where the parameter I denotes an effective moment of inertia of the nucleus, which can be calculated after assuming a specific model for the internal structure. The rotational energy levels are obtained when the angular momentum is quantized and the result for an even-even nuclei is

$$E_{rot} = \frac{\hbar^2}{2I} J(J+1) \tag{3.5}$$

where J denotes the total angular quantum number of the nucleus. In the case of a spheroidal nucleus, the deformation is symmetric with respect to the reflection in the nuclear center. Therefore, the J is restricted to even values. That is, the J can have the values 0, 2, 4, 6, ..., and thus, the parity should be even. Hence, we find that these low lying rotational levels be characterized by a sequence of the states with $J = 0^+$, 2^+, 4^+, etc. with even parity throughout and the corresponding the energies proportional to $J(J+1)$, so that E_0, E_2, E_4, E_6, E_8, ... are proportional to 0, 6, 20, 42, 72, ..., respectively. It may be seen that the ratio of excitation of successive states is

$$\frac{E_4}{E_2} = \frac{10}{3} \qquad \frac{E_6}{E_2} = 7 \qquad \frac{E_8}{E_2} = 12$$

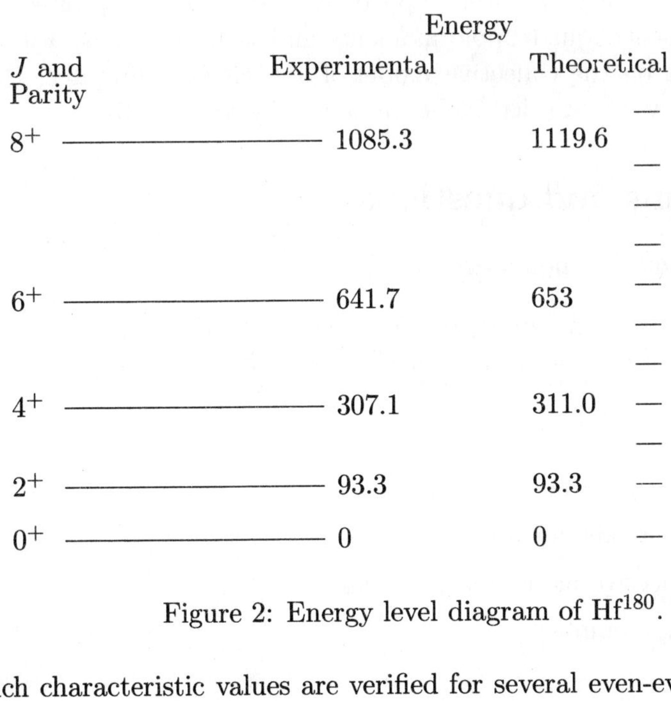

Figure 2: Energy level diagram of Hf^{180}.

Such characteristic values are verified for several even-even nuclei. For example, Figure 2 shows the energy level diagram of $_{72}Hf^{180}$.The excitation energies are measured with great precision with the help of a bent crystal spectrometer. Substituting experimental value of energy for the 2^+ level in equation (3.5), we get $\hbar^2/2I = 15.55$. Using this value, the energies of other levels are calculated and are written to the extreme right of each level in the diagram.

The agreement is found to be quite good, but there is a symmetric difference which increases with the increase of the excitation energy. This difference may however be assigned to the increase in the moment of inertia I, because of the action of the centrifugal force. After accounting for this effect, the equation (3.5) may be expressed as

$$E_{rot} = \frac{\hbar^2}{2I} J(J+1) - BJ^2(J+1)^2$$

where B is a constant. Now, the agreement with the experimental values is extremely good.

Thus, we see that the collective model is quite successful in interpreting the pattern of excited states of even-even nuclei. For odd-even nuclei, the situation is a bit complicated by the fact that the motion of the core and motion of the odd nucleons must be coupled and thus the motion of the odd nucleon is no longer adequately described by the shell model states which correspond to a static spherical core.

Finally, the theory has been applied successfully to the problems of magnetic moments, quadrupole moments and isomeric transpositions. For their work on the collective model of a nucleus, Rainwater, Bohr and Mottelson were awarded Nobel prize in Physics for 1975.

5. Problems and questions

1. Discuss about the liquid drop model.

2. Describe the magic numbers. How the magic numbers can be explained with the help of the shell model for a nucleus.

3. Write notes on the following

   (i) Liquid drop model for a nucleus

   (ii) Shell model for a nucleus

   (iii) Collective model for a nucleus

   (iv) Magic numbers

IV. Nuclear Reactions

In a nuclear reaction, in general, two nuclei interact with each other. For the convenience of theoretical treatment the bigger one of them is generally taken on rest and is known as the target. The lighter one called the projectile moves towards the target. The projectile may be a gamma radiation in place of the smaller nucleus. As a consequence of the nuclear reaction, in general, the output is in the form of two nuclei. The bigger is known as the product nucleus and the smaller as the out going nucleus. This out going nucleus may be a gamma radiation.

A nuclear reaction cannot take place spontaneously when the mass of product nuclei is larger than that of the nuclei participating in the reaction. However, such reaction may be possible when sufficient energy is supplied from out side. For example, such reaction may be produced by shooting a projectile particle towards the target with sufficiently high energy to overcome the mass difference. Such a reaction is known as the artificial nuclear reaction.

1. Kinds of nuclear reactions

Before going for the quantitative aspects of nuclear reactions, let us first consider some important mechanisms by which such reactions take place in general. Based on the criterion whether energy exchange between the projectile and target takes place or not, the reactions can be mainly classified in two categories, called the elastic scattering and inelastic scattering.

1.1 Elastic scattering

In the elastic scattering, an incident particle collides with a target nucleus and goes away without net exchange of energy. However, the direction of motion of the incident particle relative to the target changes, in general. Scattering of alpha-particles in the Rutherford experiment is a good example of this process.

$$_2He^4 + \ _{79}Au^{197} \rightarrow \ _{79}Au^{197} + \ _2He^4$$

1.2 Inelastic scattering

In an inelastic scattering, there is a net exchange of kinetic energy between the incident particle and the target nucleus. This energy may be used for exciting the nucleus to a higher quantum state. This inelastic scattering can be represented by the well known example

$$_3Li^7 + \ _1H^1 \rightarrow (_3Li^7)^* + \ _1H^1$$

The star is used to indicate that the nucleus is in an excited state. In this example, the excess energy is latter on radiated away in the form of a γ-radiation. The incident particle may be a γ photon. In the inelastic scattering completely new nuclei may be formed.

Note: The category of inelastic scattering includes several types of reactions. Some categories of reactions are as the following.

1.3 Disintegration

On striking the target nucleus, the incident particle is absorbed and a different kind of particle is ejected. The product nucleus differs from the target nucleus. The incident particle is generally a light particle, such as α-particle, proton, neutron, etc. A typical reaction of this type is as the following.

$$_2N^{14} + \ _2He^4 \rightarrow \ _8O^{17} + \ _1H^1$$

1.4 Photo disintegration

In this process, the incident particle is a γ-radiation which is absorbed by the target nucleus. The nucleus is excited to a higher quantum state. Where the energy is high enough, one or more particles may be liberated. An example of such reaction is as the following.

$$_1H^2 + \ \gamma \rightarrow \ _1H^1 + \ _0n^1$$

1.5 Radiative capture

In this process, the incident particle combines with the target nucleus to form a new nucleus (also known as a compound nucleus) which is in an excited state. The excess energy is now emitted in the form of γ photons. An example of such process is as the following.

$$_{12}\mathrm{Mg}^{26} + {}_{1}\mathrm{H}^{1} \rightarrow ({}_{13}\mathrm{Al}^{27})^{*} \rightarrow {}_{13}\mathrm{Al}^{27} + \gamma$$

The star is used to indicate that the nucleus is in an excited state.

1.6 Direct reactions

In this process, the incident particle may pull out one of the nucleons from the target nucleus (called the pickup reaction) or when the incident particle has more than one particle, it may give one or more nucleons to the target nucleus (called the stripping reaction). An example of stripping reaction is as the following.

$$_{29}\mathrm{Cu}^{63} + {}_{1}\mathrm{d}^{2} \rightarrow {}_{29}\mathrm{Cu}^{64} + {}_{1}\mathrm{H}^{1}$$

1.7 Heavy ion reactions

In these reactions, besides the target nucleus, the incident nucleus also is heavy. Here, we find the characteristics of both the compound nucleus, and the stripping and pick-up mechanisms. One example of such reaction is as the following.

$$\mathrm{N}^{14} + \mathrm{Pb}^{207} \rightarrow \mathrm{N}^{13} + \mathrm{Pb}^{208}$$

In this example, the nitrogen has lost one nucleon whereas the lead has gained one nucleon.

1.8 Spontaneous decay

A radioactive nucleus can spontaneously emit α or β particles. There is no need of nay external agent. With the emission of an α or β particle, we get a new nucleus. In the following examples, a nucleus X decays into the nucleus Y and releases one α or β particle:

$$_{Z}\mathrm{X}^{A} \rightarrow {}_{Z-2}\mathrm{Y}^{A-4} + \alpha \qquad\qquad {}_{Z}\mathrm{X}^{A} \rightarrow {}_{Z+1}\mathrm{Y}^{A} + \beta$$

With the emission of one α-particle, the atomic number decreases by two and the mass number decreases by four. With the emission of one β-particle, the atomic number increases by one whereas the mass number remains unchanged.

1.9 Spallation reactions

On capture of an incident particle, a heavy nucleus may have sufficient energy required for ejection of several particles. Such a reaction is known as the spallation reactions. For example, in the nuclear fission process a heavy nucleus U^{235} splits into two main nuclei Zr^{98} and $_{52}Te^{136}$ and two neutrons as the following.

$$_{92}U^{235} + {_0n^1} \rightarrow {_{40}Zr^{98}} + {_{52}Te^{136}} + 2\,{_0n^1}$$

1.10 High energy reactions

These are reactions where the energy of incident particle is more than few MeV. These reactions produced first time the particles, such as mesons and many more strange particles. Owing to quite high energy of incident particles, these reactions are placed under the category of high energy.

2. Conservation laws

In a nuclear reaction, certain quantities must remain conserved. Here, we shall mention in brief about the conservation laws for various quantities which appear to be valid in ordinary nuclear interactions.

2.1 Conservation of energy-mass

In classical mechanics, we discuss about the conservation of mass and conservation of energy separately. When we consider a system at the atomic dimensions, the energy and mass remain no longer independent quantities. Rather the energy and mass are exchangeable through the energy-mass equivalence relation, $\Delta E = \Delta m\, c^2$. Here, c is the speed of light. Now, according to the energy-mass conservation, the sum of the mass and energy of all particles before a reaction is equal to the sum of the mass and energy of all the products after the reaction.

2.2 Conservation of momentum

The sum of the linear momenta of all the products must be equal to the sum of the linear momenta of all the particles participating in the reaction (for the convenience of mathematical derivations, the target nucleus is generally taken at rest).

2.3 Conservation of angular momentum

The sum of the angular momenta of all the products must be equal to the sum of the angular momenta of all the particles participating in the reaction (for the convenience of mathematical derivations, the target nucleus is generally taken at rest).

2.4 Conservation of charge

The sum of the electric charges of all the products must be equal to the sum of the electric charges of all the particles participating in the reaction.

2.5 Conservation of nucleons

According to this law of conservation, the nucleons can neither be created nor destroyed. Therefore, the number of nucleons minus the number of anti-nucleons in a reaction must remain conserved.

2.6 Conservation of spin

Following the law of conservation, the spin character of a closed system does not change. That is, the statistics after a reaction remains the same as that was before the reaction. We have tow kinds of statistics: Fermi-Dirac statistics (for fermions) and Bose-Einstein statistics (for bosons).

2.7 Conservation of parity

Parity is a property of each particle. Total parity of an interacting system is the product of the intrinsic-parities of the target nucleus and the incident particle. No violation of parity has been observed in a nuclear reaction (strong nuclear forces). Although parity does not appear to be conserved in the weak interactions.

3. Nuclear reaction kinematics

From the energy point of view, nuclear reactions can be classified into two categories: (i) endothermic and (ii) exothermic. An exothermic reaction is possible even when the incident particle has no energy. Contrary to that, for an endothermic reaction to take place, a minimum amount of energy of the incident nucleus is required. Such restriction may be understood with the help of the laws of conservation of energy-mass and of momentum. These restrictions are known as the kinematic restrictions and the mathematical method is known as the kinematics. Let us consider a nuclear reaction:

$$X + x \rightarrow Y + y$$

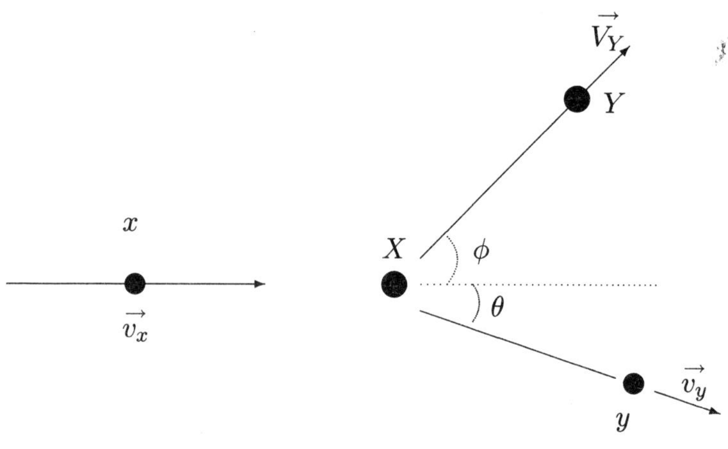

Figure 1: A nucleus x reacts with another nucleus X. After reaction, the nucleus y moves in a direction θ and the nucleus Y moves in a direction ϕ as shown in the figure. For convenience, the target nucleus X is taken at rest.

where, X, x, Y and y are the target nucleus, incident particle, product nucleus and the outgoing particle, respectively. For convenience, we take the target nucleus X to be at rest and thus its kinetic energy is zero (Figure 1). Following the law of energy-mass conservation for the reaction, we get

$$M_X c^2 + (E_x + m_x c^2) = (E_Y + M_Y c^2) + (E_y + m_y c^2) \tag{4.1}$$

where, M_X, m_x, M_Y and m_y are the masses of X, x, Y and y, respectively. The E_x, E_Y and E_y are the kinetic energies of x, Y and y, respectively. We now introduce a parameter Q, which represents the difference between the kinetic energy of the products of the reaction and the kinetic energy of the incident particle as

$$Q = E_Y + E_y - E_x \tag{4.2}$$

From equations (4.1) and (4.2), we have

$$(M_X + m_x - M_Y - m_y)c^2 = Q$$

The quantity Q is generally known as the Q-value of the reaction. When the Q is positive, the reaction is said to be 'exoergic' or 'exothermic'.

The exothermic reaction takes place when the sum of the masses of particles before reaction is larger than the masses of particles after the reaction. For such reactions, the incident nucleus need not to have any kinetic energy. Following the law of energy-mass conservation, the sum of the kinetic energies of the particles after the reaction is greater than that of the particles before the reaction.

When the Q is negative, the reaction is said to be 'endoergic' or 'endothermic'. For such reactions, this much amount of energy is supplied, which is generally in the from of the kinetic energy of incident particle.

The term E_Y in equation (4.1) represents the recoil energy of the product nucleus. It is usually small and difficult to measure, but can be eliminated by taking into account the conservation of momentum. In an experiment, the Q-value can be measured with the help of the incident energy E_x and the energy E_y of the ejected particle at some specified angle θ. On applying the laws of conservation of momentum, we have

$$m_x v_x = M_Y V_Y \cos\phi + m_y v_y \cos\theta \tag{4.3}$$

and

$$M_Y V_Y \sin\phi = m_y v_y \sin\theta \tag{4.4}$$

where v_x, v_y and V_Y are the velocities of x, y and Y, respectively. After eliminating ϕ from equations (4.3) and (4.4), we have

$$M_Y^2 V_Y^2 = m_x^2 v_x^2 + m_y^2 v_y^2 - 2m_x m_y v_x v_y \cos\theta \tag{4.5}$$

We know that

$$E_x = \frac{1}{2}\, m_x v_x^2, \qquad E_y = \frac{1}{2}\, m_y v_y^2, \qquad E_Y = \frac{1}{2}\, M_Y V_Y^2 \tag{4.6}$$

After eliminating v_x, v_y and V_Y in equation (4.5) with the help of equation (4.6), we get

$$2E_Y M_Y = 2E_x m_x + 2E_y m_y - 4(m_x m_y E_x E_y)^{1/2} \cos\theta$$

or

$$E_Y = E_x\, \frac{m_x}{M_Y} + E_y\, \frac{m_y}{M_Y} - \frac{2}{M_Y}\, (m_x m_y E_x E_y)^{1/2} \cos\theta$$

Substituting this value of E_Y in equation (4.2), we get

$$Q = E_x\, \frac{m_x - M_Y}{M_Y} + E_y\, \frac{m_y + M_Y}{M_Y} - \frac{2}{M_Y}\, (m_x m_y E_x E_y)^{1/2} \cos\theta \tag{4.7}$$

Equation (4.7) is known as the Q-equation of the reaction. It provides the desired relation between the energy released and the measured quantities E_x, E_y and θ. For a special case, when we are investigating the out coming particle y at 90° to a collimated beam of projectile, equation (4.7) reduces to

$$Q = E_y\left(1 + \frac{m_y}{M_Y}\right) - E_x\left(1 - \frac{m_x}{M_Y}\right) \tag{4.8}$$

With the help of the modern techniques, it is possible to measure Q-values for nuclear reactions to a good accuracy.

3.1 General solution of Q-equation

For obtaining the variation of E_y with respect to E_x, for a given value of Q, equation (4.7) can be expressed as

$$E_y - 2u\sqrt{E_y} - v = 0 \tag{4.9}$$

where

$$u = \frac{\sqrt{(m_x m_y E_x)}}{M_Y + m_y}\cos\theta \qquad \text{and} \qquad v = \frac{M_Y Q + E_x(M_Y - m_x)}{M_Y + m_y}$$

Equation (4.9) is a quadratic equation in $\sqrt{E_y}$ and its solution is

$$\sqrt{E_y} = u \pm \sqrt{u^2 + v} \tag{4.10}$$

The nuclear reactions are divided energetically into two categories: (i) Exothermic reactions and (ii) endothermic reactions. These two cases will be discussed here.

(i) Exothermic reactions: For these reactions Q is always positive. Such reactions are possible even when $E_x = 0$, *i.e.*, even when the particles before the reaction have no kinetic energy.

(a) For $E_x \to 0$, we have $u = 0$ and $v = M_Y Q/(M_Y + m_y)$. Thus, equation (4.10) gives

$$E_y = QM_Y/(m_y + M_Y)$$

Since $Q > 0$, the energy of the particle y is always positive and does not depend on the angle θ.

(b) When $\theta + \phi = \pi$, from equation (4.4), we have

$$M_Y V_Y = m_y v_y; \qquad M_Y\sqrt{\frac{2E_Y}{M_Y}} = m_y\sqrt{\frac{2E_y}{m_y}}; \qquad M_Y E_Y = m_y E_y$$

Here, we have used equation (4.6). Thus, for $E_x \to 0$, from equation (4.8), we have

$$Q = E_y \left(1 + \frac{m_y}{M_Y}\right) = E_y + E_Y$$

(c) Generally, the projectile x is lighter than the product nucleus Y, so that $m_x < M_Y$. Thus, v is positive for all values of the incident energy E_x and therefore only one solution $\sqrt{E_y} = u + \sqrt{u^2 + v}$ is positive. Thus, E_y is single valued when $Q > 0$ and $M_Y > m_x$. Further, E_y has the lowest value when $\theta = 180^o$, *i.e.*, in the backward direction.

(ii) Endothermic reactions: For these reactions, Q is negative. For $Q < 0$, when $E_x \to 0$, we have $u = 0$ and $v < 0$. Thus, $\sqrt{E_y}$ is not real, showing that such reactions are not possible when $E_x = 0$. It shows that the endothermic reactions are only possible when E_x is sufficiently large. The smallest value of E_x at which the reaction can take place is known as the 'threshold energy'. At the threshold energy, we have $u^2 + v = 0$, so that

$$\frac{(m_x m_y E_x)}{(M_Y + m_y)^2} \cos^2\theta + \frac{M_Y Q + E_x(M_Y - m_x)}{(M_Y + m_y)} = 0$$

This equation can be rearranged as

$$E_x\left[\frac{m_x m_y \cos^2\theta}{(M_Y + m_y)^2} + \frac{(M_Y - m_x)}{(M_Y + m_y)}\right] + \frac{M_Y Q}{(M_Y + m_y)} = 0$$

Thus, we have

$$E_x = -Q\left[\frac{m_y + M_Y}{m_y + M_Y - m_x - (m_x m_y / M_Y)\sin^2\theta}\right]$$

Since $Q < 0$, we have $E_x > 0$. The E_x depends on the angle θ and will become the minimum when $\theta = 0$. Thus, the threshold energy, *i.e.*, the minimum energy required for the reaction to take place is

$$(E_x)_{th} = -Q\left[\frac{M_Y + m_y}{M_Y + m_y - m_x}\right]$$

4. Compound nucleus model

For the study of nuclear reactions, several models for reaction mechanism have been formulated from time to time. In the present discussion, we shall consider: (i) compound nucleus model and (ii) the optical potential model, in brief.

In 1936, Bohr proposed his theory of compound nucleus, which has been found extremely useful for correlation and interpretation of nuclear reactions. According to this theory, a nuclear reaction takes place in two distinct and independent steps:

(i) The incident particle is absorbed by the target nucleus to form a compound nucleus:

incident particle + target nucleus → compound nucleus

(ii) The compound nucleus so formed disintegrates by ejecting a particle and leaving the product nucleus.

compound nucleus → product nucleus + outgoing particle

Bohr assumed also that the mode of disintegration of the compound nucleus is independent of the way in which it was formed, and it depends only on its properties, such as its energy and angular momentum.

The first step about the formation of a compound nucleus can be expressed as:

$$x + X \rightarrow (C)^*$$

Here, x is an incident particle, such as proton, neutron, α-particle, etc. or a γ-photon. It is captured by the target nucleus X to form a compound nucleus C. The star on the nucleus C denotes that the nucleus is in the excited state. As soon as the incident particle x enters into the nucleus X, the energy of x is quickly shared among the particles inside X. On being captured, the incident particle x makes available a certain amount of excitation energy, which is nearly equal to the kinetic energy of the captured particle plus its binding energy in the compound nucleus. The magnitude of the excitation energy can be calculated with the help of the masses of the incident particle, target nucleus and the compound nucleus, and the kinetic energy of the incident particle. Let us consider an example of capture of neutron by ${}_{13}\mathrm{Al}^{27}$ to form a compound nucleus $({}_{13}\mathrm{Al}^{28})^*$:

$${}_{13}\mathrm{Al}^{27} + {}_{0}\mathrm{n}^{1} \rightarrow ({}_{13}\mathrm{Al}^{28})^*$$

The masses of the interacting neutron and the aluminum nucleus are 1.00898 and 26.99008 amu or a total of 27.99906 amu. However, the mass of the compound nucleus is 27.99077 amu. The mass excess is 0.00829 amu which corresponds a kinetic energy of 7.72 MeV. It should obviously be the kinetic energy of the incident neutron.

When the excitation energy is large enough, a single particle or a group of nucleons may escape, and the compound nucleus disintegrates into the product nucleus and a outgoing particle. The energy must be concentrated on a single particle or a group of nucleons in order to separate it from the compound nucleus:

$$(\mathrm{C})^* \rightarrow \mathrm{Y} + \mathrm{y}$$

The compound nucleus $(\mathrm{C})^*$ ejects the particle y, leaving the product nucleus Y. Hence, the complete process may be expressed as

$$\mathrm{x} + \mathrm{X} \rightarrow (\mathrm{C})^* \rightarrow \mathrm{Y} + \mathrm{y}$$

This reaction is denoted as $X(x, y)Y$ or simply (x, y).

4.1 Virtual levels

There are several ways in which the excitation energy of the compound nucleus can be distributed among the nucleons and since each distribution corresponds to a virtual level, there are many possible virtual levels of the compound nucleus. These levels are closely related to the phenomenon of resonance. It is reasonable to assume that when the energy of incident particle is such that the total energy of the incident particle plus target nucleus is equal to the excitation energy of one of these levels of the compound nucleus, the probability of formation of compound nucleus becomes much larger than that of the situation when the energy falls between two successive levels.

When energy of incident particle is varied, one finds a resonance peak in the rate of a nuclear reaction, showing that the compound nucleus has an energy level whose excitation energy is very close to the sum of the binding energy of that particle and its kinetic energy. For example, for the reaction $\mathrm{Al}^{27}(\alpha, p)\mathrm{Si}^{30}$, 18 resonances have been observed for the energies of α particle in the range from 3.95 MeV to 8.62 MeV. The excitation energy of a level cannot be obtained just by adding the resonance energy to the binding energy, as it would neglect the motion of the compound nucleus.

The fraction of particle energy that must be added to the binding energy in order to get the excitation energy can be obtained in the following manner. When the target nucleus is assumed to be at rest before the collision, the momentum of the incident particle must be equal to that of the compound nucleus:

$$m_x v_x = M_{CN} V \qquad \text{and therefore} \qquad V = \frac{m_x}{M_{CN}}\, v_x$$

where v_x, V and M_{CN} are the speed of the incident particle, and the speed and mass of the compound nucleus, respectively. The fraction of energy of the incident particle that goes into the excitation of the initial nucleus, denoted as E'_x, is expressed as

$$E'_x = \frac{1}{2}\, m_x v_x^2 - \frac{1}{2}\, M_{CN} V^2 = \frac{1}{2}\, m_x v_x^2 - \frac{1}{2}\, M_{CN} v_x^2\, \frac{m_x^2}{M_{CN}^2}$$

$$= \frac{1}{2}\, m_x v_x^2 \Big(1 - \frac{m_x}{M_{CN}}\Big) = E_x \Big(1 - \frac{m_x}{M_{CN}}\Big)$$

Here, $E_x = m_x v_x^2/2$ is the energy of the incident particle. Using $M_{CN} = m_x + M_X$, this equation may be expressed as

$$E'_x = E_x \Big(\frac{M_X}{m_x + M_X}\Big)$$

Each of the excited states of the compound nucleus, whether bound or virtual, has a certain mean life-time τ. That is, there is a certain period of time, on the average, during which the nucleus remains in a given excited state before it decays by emitting either a particle or a γ-radiation. The reciprocal of the mean life-time is known as the disintegration constant, which gives the probability per unit time of the emission of a particle or a γ-radiation. In the discussion of energy states excited by the nuclear reactions, we use the term 'level width', denoted by Γ and expressed as

$$\Gamma = \frac{h}{2\pi\tau}$$

The level width has the unit of energy. For each possible mode of decay, there is different probability and therefore, a different partial width Γ_y for each decay product. The total width Γ of an energy level is than the sum of the individual partial widths:

$$\Gamma = \Gamma_\gamma + \Gamma_\alpha + \Gamma_p + \ldots$$

where Γ_γ is a measure of the probability per unit time for emission of a γ-radiation, Γ_α the α-particle width, Γ_p the proton width, and so on. The concept of level widths is useful as the values of these widths can often be obtained from the measurement of resonances. When the total width is known the value of life-time is

$$\tau(\mathrm{s}) = \frac{h}{2\pi\Gamma} = \frac{6.62 \times 10^{-34}}{2\pi\Gamma(\mathrm{eV}) \times 1.60 \times 10^{-19}} = \frac{6.6 \times 10^{-16}}{\Gamma(\mathrm{eV})}$$

Level widths are generally expressed in eV.

5. Cross sections for nuclear reactions

For the study of a nuclear reaction in detail, it is necessary to have a quantitative measure of probability of the reaction. This quantity needs to be measured experimentally and could be calculated in such a way that the theoretical and experimental values could be compared easily. The quantity which is commonly used for this purpose is known as the cross-section of a nucleus for a particular reaction and is usually denoted by σ along with an appropriate subscript. The idea of a nuclear cross-section may be considered as the cross sectional area, or the target area, presented by a nucleus to an incident particle. When the nuclei are considered as spheres of radius R and the incident particles as a point-projectiles, then the target area, or the cross section σ of each nucleus is $\sigma = \pi R^2$.

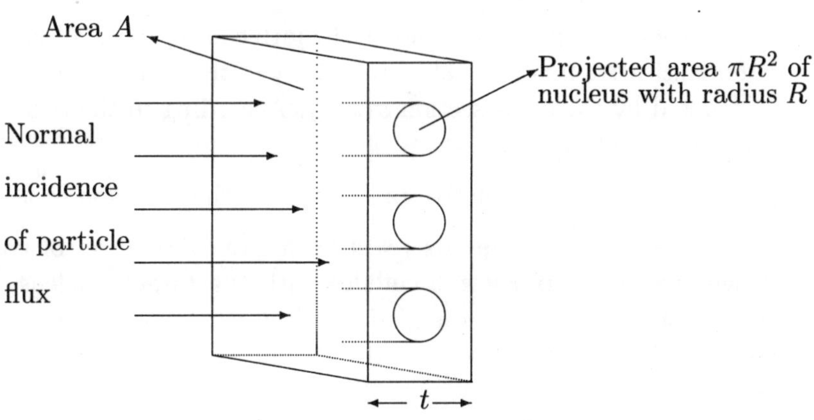

Figure 2: Cross section for nuclear reaction as geometrical areas.

A particle which incident normally on a thin sheet of material of area A and having N_T nuclei has a probability $N_T\sigma/A$ of colliding with a nucleus provided that there is no overlapping of the nuclei (Figure 2). The quantity N_T/A, which is the number of nuclei per unit area or the surface density of nuclei, is equal to Nt, where N is the number of nuclei per unit volume and t is the thickness of the sheet. Suppose the incident beam contains n particles per unit volume and is moving with a velocity v, then the number of particles passing through the sheet is nv per unit area per second. Now, the collision rate can be expressed as

$$\text{collision per unit area per second} = nv\,\frac{N_T\sigma}{A} = nv\sigma Nt$$

The collision cross section can be expressed as

$$\sigma = \frac{\text{collision per unit area per second}}{nvNt}$$

The quantity nv, which is the number of particles in the incident beam crossing unit area each second, is known as the particle flux.

The above discussion is applicable to any kind of nuclear reaction, and the term 'collision' can be replaced by the corresponding term. For example, σ_s may denote the cross section for scattering of a given particle, and may have two parts: σ_e, the cross section for elastic scattering and σ_i, the cross section for inelastic scattering. There can be cross sections for individual reaction, such as $\sigma(\alpha, p)$, $\sigma(\alpha, n)$ and $\sigma(p, \alpha)$.

6. Breit-Wigner formula

The concept of cross section and level width can be applied to resonances in a quantitative manner. The probability of the reaction $X(x, y)Y$ may be denoted by the cross section $\sigma(x, y)$. According to the two-step view of nuclear reactions,

$$\sigma(x, y) = \sigma_c(x) \times \text{ (relative probability of the emission of } y)$$

where $\sigma_c(x)$ denotes the cross section for the formation of compound nucleus when the particle x collides with the target nucleus X. The relative probability for the emission of the particle y is given by Γ_y/Γ, where Γ_y is the partial level-width for the emission of y and Γ is the total level-width.

In general, the values of the cross-sections as well as of the level-widths depend on the energy of incident particle as well as on the charge and mass of the target nucleus. Here, one of the problems of the theory is the calculation of $\sigma_c(x)$, Γ, Γ_y, and therefore the calculation of $\sigma(x, y)$. For the cases of particular importance, in the resonance processes, a theoretical formula was derived by Breit and Wigner. The rigorous derivation of that formula is a tedious job and is out of the scope of this presentation. However, the important features of the formula can be described in a qualitative manner. In its simplest form, the Breit-Wigner formula gives the value of cross section in the neighbourhood of a single resonance level formed by an incident particle with zero angular momentum ($l = 0$ partial wave). Under these conditions, the formula can be expressed as

$$\sigma(x, y) = \frac{\lambda^2}{4\pi} \frac{\Gamma_x \Gamma_y}{(E - E_0)^2 + (\Gamma/2)^2} \tag{4.11}$$

where λ is the de Broglie wavelength of the incident particle, E the energy, E_0 the energy at the peak of the resonance and the *Gamma*'s the widths. This expression may be considered as consisting of three factors. The first one is the measure of the probability of formation of a compound nucleus, which is proportional to λ^2 according to the wave mechanics. The second factor

$$\frac{1}{(E - E_0)^2 + (\Gamma/2)^2}$$

describes the resonance property and may be called the 'resonance factor'. The value of the denominator of this factor becomes the smallest when $E = E_0$ and then the cross section becomes the largest. As the value of E departs from E_0, the value of the denominator increases and the resonance factor as well as the cross section decrease. The third factor is the probability for definite types of disintegration of the compound nucleus and is expressed through the partial widths Γ_x and Γ_y.

The Breit-Wigner formula is an example of application of wave mechanics to nuclear physics, as the wave length of incident particle appears there. Values of $\sigma_0(x, y)$ (the cross section at the peak of the resonance), E_0 and Γ can be obtained from the experiments where the transmission of particle through the target is measured as a function of energy of the incident particles in the neighbourhood of the resonance energy. Equation (4.11) shows that the full-width at the half-maximum, *i.e.*, the energy interval between the two points at which the cross-section is $\sigma_0/2$, is equal to the total width σ. Thus, the total width can be determined directly from the actual observed width of the cross-section resonance. After knowing the total width, the values of partial widths can be obtained by further experiments, or may some times be inferred with the help of certain theoretical considerations.

Exercise 1. With the help of the Breit-Wigner formula

$$\sigma(x, y) = \frac{\lambda^2}{4\pi} \frac{\Gamma_x \Gamma_y}{(E - E_0)^2 + (\Gamma/2)^2}$$

show that the full-width at half-maximum is equal to the total width Γ.

Solution: The cross-section is the maximum $\sigma_0(x, y)$ when $E = E_0$. Thus, we have

$$\sigma_0(x, y) = \frac{\lambda^2}{4\pi} \frac{\Gamma_x \Gamma_y}{(\Gamma/2)^2}$$

Suppose, we have half-maximum $\sigma_0/2$ when $E = E'$. Then we have

$$\frac{\sigma_0}{2} = \frac{\lambda^2}{4\pi} \frac{\Gamma_x \Gamma_y}{(E' - E_0)^2 + (\Gamma/2)^2}$$

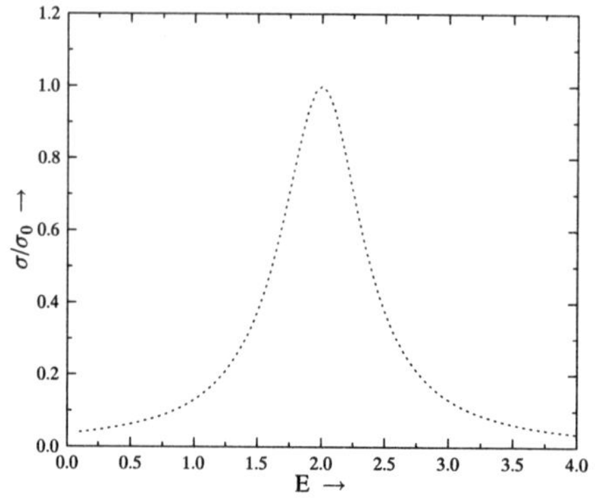

Using the value of σ_0 here, we get

$$\frac{1}{2}\frac{\lambda^2}{4\pi}\frac{\Gamma_x\Gamma_y}{(\Gamma/2)^2} = \frac{\lambda^2}{4\pi}\frac{\Gamma_x\Gamma_y}{(E'-E_0)^2+(\Gamma/2)^2}$$

It gives

$$(E'-E_0)^2+(\Gamma/2)^2 = 2(\Gamma/2)^2 \qquad \text{or} \qquad (E'-E_0) = \pm\Gamma/2$$

Thus, the full width at half-maximum is

$$2|(E'-E_0)| = \Gamma$$

It obviously shows that the full width at half-maximum is equal to the total width Γ.

Exercise 2. With the help of the Breit-Wigner formula

$$\sigma(x,y) = \frac{\lambda^2}{4\pi}\frac{\Gamma_x\Gamma_y}{(E-E_0)^2+(\Gamma/2)^2}$$

obtain the full width at quarter-maximum.

Solution: The cross-section is the maximum $\sigma_0(x,y)$ when $E = E_0$. Thus, we have

$$\sigma_0(x,y) = \frac{\lambda^2}{4\pi}\frac{\Gamma_x\Gamma_y}{(\Gamma/2)^2}$$

Suppose, we have quarter-maximum $\sigma_0/4$ when $E = E'$. Then we have

$$\frac{\sigma_0}{4} = \frac{\lambda^2}{4\pi}\frac{\Gamma_x\Gamma_y}{(E'-E_0)^2+(\Gamma/2)^2}$$

Using the value of σ_0 here, we get

$$\frac{1}{4}\frac{\lambda^2}{4\pi}\frac{\Gamma_x\Gamma_y}{(\Gamma/2)^2} = \frac{\lambda^2}{4\pi}\frac{\Gamma_x\Gamma_y}{(E'-E_0)^2+(\Gamma/2)^2}$$

It gives

$$(E'-E_0)^2+(\Gamma/2)^2 = 4(\Gamma/2)^2 \qquad \text{or} \qquad (E'-E_0) = \pm\sqrt{3}\,\Gamma/2$$

Thus, the full width at quarter-maximum is

$$2|(E'-E_0)| = \sqrt{3}\,\Gamma$$

Thus, the full width at quarter-maximum is $\sqrt{3}$ times the total width Γ.

7. Optical model

In the optical model, the nucleus is replaced by a complex potential having both real and imaginary parts. This optical potential is used in the Schrödinger equation, which is solved with the help of the partial wave analysis. The real part of the potential gives the elastic scattering whereas the imaginary part gives the sum of the inelastic and reaction cross sections.

Let us develop an optical potential which can give which can reproduce the experimentally observed cross sections when the incident particle is either a neutron or a proton. For any other projectile, minor changes need to be made in the optical potential.

A commonly used optical potential consists of a central potential with real and imaginary parts and a spin-orbit potential with real and imaginary parts. When the incident particle has charge, an additional Coulomb potential is included. Therefore, the potential is

$$V(r) = -V_0 f(r) - iW_0 g(r) + \left(\frac{\hbar}{m_\pi c}\right)^2 (V_{SO}+iW_{SO})\left(\frac{1}{r}\frac{\mathrm{d}f}{\mathrm{d}r}\right)\vec{\sigma}\,.\,\vec{l} + V_c(r)$$

There are a large number of parameters, V_0, W_0, V_{SO}, W_{SO} and some form factors $f(r)$, $g(r)$, $V_c(r)$. These parameters and form factors are adjusted so that the experimental cross sections are reproduced. The quantity $\hbar/m_\pi c$ is the Compton wavelength of the pion which is approximately 1.4 fm. The form factor $f(r)$ is generally taken in the form of Fermi distribution function:

$$f(r) = \left[1+\exp\left(\frac{r-R}{a}\right)\right]^{-1}$$

where $R = 1.25A^{1/3}$ and $a = 0.65$ fm. The parameter R represents the potential well radius which is taken slightly larger than the nuclear charge radius and a is the surface thickness. For the absorption-form factor $g(r)$, several choices have been used. One of them describing the volume absorption is expressed as

$$g(r) = f(r) = \left[1 + \exp\left(\frac{r-R}{a}\right)\right]^{-1}$$

The other choice describing the surface absorption is as the following.

$$g(r) = \exp\left[-\frac{(r-R)^2}{b^2}\right]$$

with

$$b = \begin{cases} 0.9 \text{ fm} & \text{for neutrons} \\ 1.2 \text{ fm} & \text{for protons} \end{cases}$$

One more choice in use has been as the following.

$$g(r) = -\frac{df}{dr}$$

The parameters V_0, W_0, V_{SO} and W_{SO} vary smoothly with the incident neutron energy. For the variation of the energy of incident neutron from from 3 MeV to 20 MeV, the parameter V_0 decreases from 40 MeV to 10 MeV, the parameter W_0 increase from 3 MeV to 20 MeV, the spin-orbit coupling parameter V_{SO} decreases from 10 MeV to about 1 MeV,and the parameter W_{SO} remains almost zero. If the incident particle is proton, there is an additional Coulomb term expressed as the following.

$$V_c(r) = \begin{cases} (zZe^2/2R_c)(3 - r^2/R_c^2) & r < R_c \\ zZe^2/r & r > R_c \end{cases}$$

where z and Z are, respectively, the charges of incident particle and the target nucleus, and R_c is the charge radius of the nucleus, given by ($R_c = 1.2A^{1/3}$). Here, the target nucleus is considered as a uniformly charged sphere of radius R_c.

Using this optical potential in the radial wave equation, we obtain cross sections for the elastic scattering as well as for the reaction. This equation is solved numerically with the help of computers.

The optical model method is used widely in the energy range from 10 MeV to 100 MeV, but at higher energies, the method becomes cumbersome as the contribution of higher partial waves becomes significant.

The method of optical model can be extended to study the pion-nucleus scattering. For this, we use the pion-nucleus optical potential.

8. Nuclear transmutation

The process in which a nucleus converts into another nucleus is known as the transmutation. When the energy of the incident particle is not too high (< 50 MeV), in general, a compound nucleus is formed. This compound nucleus then breaks into a product nucleus and a particle is ejected out. Here, we shall discuss some types of nuclear transmutation.

8.1 Transmutation by alpha particles

In such transmutation, an alpha particle is absorbed by the target nucleus, forming a compound nucleus. In the process of disintegration of this compound nucleus, the outgoing particle may be a proton, neutron, gamma radiation or other particle.

(a) (α, p) reaction: In such reactions, the outgoing particle is a proton. Some examples of this category are as the following.

$$
\begin{aligned}
{}_5\mathrm{B}^{10} + {}_2\mathrm{He}^4 &\rightarrow ({}_7\mathrm{N}^{14})^* \rightarrow {}_6\mathrm{C}^{13} + {}_1\mathrm{H}^1 \\
{}_9\mathrm{F}^{19} + {}_2\mathrm{He}^4 &\rightarrow ({}_{11}\mathrm{Na}^{23})^* \rightarrow {}_{10}\mathrm{Ne}^{22} + {}_1\mathrm{H}^1 \\
{}_{13}\mathrm{Al}^{27} + {}_2\mathrm{He}^4 &\rightarrow ({}_{15}\mathrm{P}^{31})^* \rightarrow {}_{14}\mathrm{Si}^{30} + {}_1\mathrm{H}^1 \\
{}_{14}\mathrm{Si}^{28} + {}_2\mathrm{He}^4 &\rightarrow ({}_{16}\mathrm{S}^{32})^* \rightarrow {}_{15}\mathrm{P}^{31} + {}_1\mathrm{H}^1 \\
{}_{16}\mathrm{S}^{32} + {}_2\mathrm{He}^4 &\rightarrow ({}_{18}\mathrm{A}^{36})^* \rightarrow {}_{17}\mathrm{Cl}^{35} + {}_1\mathrm{H}^1 \\
{}_{19}\mathrm{K}^{39} + {}_2\mathrm{He}^4 &\rightarrow ({}_{21}\mathrm{Sc}^{43})^* \rightarrow {}_{20}\mathrm{Ca}^{42} + {}_1\mathrm{H}^1 \\
{}_{21}\mathrm{Sc}^{45} + {}_2\mathrm{He}^4 &\rightarrow ({}_{23}\mathrm{V}^{49})^* \rightarrow {}_{22}\mathrm{Ti}^{48} + {}_1\mathrm{H}^1
\end{aligned}
$$

(b) (α, n) reaction: In such reactions, the outgoing particle is a proton. Some examples of this category are as the following.

$$
\begin{aligned}
{}_3\mathrm{Li}^{7} + {}_2\mathrm{He}^4 &\rightarrow ({}_5\mathrm{B}^{11})^* \rightarrow {}_5\mathrm{B}^{10} + {}_0n^1 \\
{}_4\mathrm{Be}^{9} + {}_2\mathrm{He}^4 &\rightarrow ({}_6\mathrm{C}^{13})^* \rightarrow {}_6\mathrm{C}^{12} + {}_0n^1 \\
{}_5\mathrm{B}^{11} + {}_2\mathrm{He}^4 &\rightarrow ({}_7\mathrm{N}^{15})^* \rightarrow {}_7\mathrm{N}^{14} + {}_0n^1 \\
{}_7\mathrm{N}^{14} + {}_2\mathrm{He}^4 &\rightarrow ({}_9\mathrm{F}^{18})^* \rightarrow {}_9\mathrm{F}^{17} + {}_0n^1 \\
{}_9\mathrm{F}^{19} + {}_2\mathrm{He}^4 &\rightarrow ({}_{11}\mathrm{Na}^{23})^* \rightarrow {}_{11}\mathrm{Na}^{22} + {}_0n^1 \\
{}_{13}\mathrm{Al}^{27} + {}_2\mathrm{He}^4 &\rightarrow ({}_{15}\mathrm{P}^{31})^* \rightarrow {}_{15}\mathrm{P}^{30} + {}_0n^1
\end{aligned}
$$

(c) Radiative capture: In such reactions, a gamma radiation comes out of the compound nucleus. An example of this category are as the following.

$$
{}_3\mathrm{Li}^{7} + {}_2\mathrm{He}^4 \rightarrow ({}_5\mathrm{B}^{11})^* \rightarrow {}_5\mathrm{B}^{11} + h\nu
$$

(d) When the energy of the incident α-particle is large enough, two or more nucleons may be emitted out from the compound nucleus. Some examples of such processes are: $(\alpha, 2n)$, (α, np), $(\alpha, 3n)$, $(\alpha, 4n)$, $(\alpha, 3np)$, etc. processes.

8.2 Transmutation by protons

In such transmutation a proton is absorbed by the target nucleus, forming a compound nucleus. In the process of disintegration of this compound nucleus, the outgoing particle may be an alpha particle, neutron, deuteron, gamma radiation or other particle.

(a) (p, α) reaction: In such reactions, the outgoing particle is an alpha particle. Some examples of this category are as the following.

$$\begin{array}{rcccl} {}_3\mathrm{Li}^7 + {}_1\mathrm{H}^1 & \rightarrow & ({}_4\mathrm{Be}^8)^* & \rightarrow & {}_2\mathrm{He}^4 + {}_2\mathrm{He}^4 \\ {}_3\mathrm{Li}^6 + {}_1\mathrm{H}^1 & \rightarrow & ({}_4\mathrm{Be}^7)^* & \rightarrow & {}_2\mathrm{He}^3 + {}_2\mathrm{He}^4 \\ {}_4\mathrm{Be}^9 + {}_1\mathrm{H}^1 & \rightarrow & ({}_5\mathrm{B}^{10})^* & \rightarrow & {}_3\mathrm{Li}^6 + {}_2\mathrm{He}^4 \\ {}_5\mathrm{B}^{11} + {}_1\mathrm{H}^1 & \rightarrow & ({}_6\mathrm{C}^{12})^* & \rightarrow & {}_4\mathrm{Be}^8 + {}_2\mathrm{He}^4 \\ {}_9\mathrm{F}^{19} + {}_1\mathrm{H}^1 & \rightarrow & ({}_{10}\mathrm{Ne}^{20})^* & \rightarrow & {}_8\mathrm{O}^{16} + {}_2\mathrm{He}^4 \\ {}_{13}\mathrm{Na}^{23} + {}_1\mathrm{H}^1 & \rightarrow & ({}_{12}\mathrm{Mg}^{24})^* & \rightarrow & {}_{10}\mathrm{Ne}^{20} + {}_2\mathrm{He}^4 \\ {}_{17}\mathrm{Cl}^{35} + {}_1\mathrm{H}^1 & \rightarrow & ({}_{18}\mathrm{A}^{36})^* & \rightarrow & {}_{16}\mathrm{S}^{32} + {}_2\mathrm{He}^4 \end{array}$$

(b) (p, n) reaction: In such reactions, the outgoing particle is a neutron. Some examples of this category are as the following.

$$\begin{array}{rcccl} {}_5\mathrm{B}^{11} + {}_1\mathrm{H}^1 & \rightarrow & ({}_6\mathrm{C}^{12})^* & \rightarrow & {}_6\mathrm{C}^{11} + {}_0n^1 \\ {}_8\mathrm{O}^{18} + {}_1\mathrm{H}^1 & \rightarrow & ({}_9\mathrm{F}^{19})^* & \rightarrow & {}_9\mathrm{F}^{18} + {}_0n^1 \\ {}_{11}\mathrm{Na}^{23} + {}_1\mathrm{H}^1 & \rightarrow & ({}_{12}\mathrm{Mg}^{24})^* & \rightarrow & {}_{12}\mathrm{Mg}^{23} + {}_0n^1 \\ {}_{28}\mathrm{Ni}^{58} + {}_1\mathrm{H}^1 & \rightarrow & ({}_{29}\mathrm{Cu}^{59})^* & \rightarrow & {}_{29}\mathrm{Cu}^{58} + {}_0n^1 \\ {}_{29}\mathrm{Cu}^{65} + {}_1\mathrm{H}^1 & \rightarrow & ({}_{30}\mathrm{Zn}^{66})^* & \rightarrow & {}_{30}\mathrm{Zn}^{65} + {}_0n^1 \end{array}$$

(c) (p, d) reaction: In such reactions, the outgoing particle is a deuteron. Some examples of this category are as the following.

$$\begin{array}{rcl} {}_3\mathrm{Li}^7 + {}_1\mathrm{H}^1 & \rightarrow & {}_3\mathrm{Li}^6 + {}_1\mathrm{H}^2 \\ {}_4\mathrm{Be}^9 + {}_1\mathrm{H}^1 & \rightarrow & {}_4\mathrm{Be}^8 + {}_1\mathrm{H}^2 \end{array}$$

(d) Radiative capture: In such reactions, a gamma radiation comes out of the compound nucleus. An example of this category are as the following.

$$\begin{array}{rcccl} {}_3\mathrm{Li}^7 + {}_1\mathrm{H}^1 & \rightarrow & ({}_4\mathrm{Be}^8)^* & \rightarrow & {}_4\mathrm{Be}^8 + \gamma \\ {}_6\mathrm{C}^{12} + {}_1\mathrm{H}^1 & \rightarrow & ({}_7\mathrm{N}^{13})^* & \rightarrow & {}_7\mathrm{N}^{13} + \gamma \\ {}_7\mathrm{N}^{14} + {}_1\mathrm{H}^1 & \rightarrow & ({}_8\mathrm{O}^{15})^* & \rightarrow & {}_8\mathrm{O}^{15} + \gamma \\ {}_9\mathrm{F}^{19} + {}_1\mathrm{H}^1 & \rightarrow & ({}_{10}\mathrm{Ne}^{20})^* & \rightarrow & {}_{10}\mathrm{Ne}^{20} + \gamma \\ {}_{13}\mathrm{Al}^{27} + {}_1\mathrm{H}^1 & \rightarrow & ({}_{14}\mathrm{Si}^{28})^* & \rightarrow & {}_{14}\mathrm{Be}^{28} + \gamma \\ {}_{24}\mathrm{Cr}^{50} + {}_1\mathrm{H}^1 & \rightarrow & ({}_{25}\mathrm{Mn}^{51})^* & \rightarrow & {}_{25}\mathrm{Mn}^{51} + \gamma \end{array}$$

(e) When the energy of incident proton is more than 20 MeV, two or more nucleons may be emitted out of the compound nucleus. Some examples are: $\mathrm{Cu}^{63}(p, pn)\ \mathrm{Cu}^{62}$, $\mathrm{Cu}^{63}(p, 2d)\ \mathrm{Zn}^{62}$ and $\mathrm{Sr}^{88}(p, 3n)\ \mathrm{Y}^{86}$

8.3 Transmutation by neutrons

In such transmutation a neutron is absorbed by the target nucleus, forming a compound nucleus. In the process of disintegration of this compound nucleus, the outgoing particle may be an alpha particle, proton, deuteron, gamma radiation or other particle.

(a) (n, α) reaction: In such reactions, the outgoing particle is an alpha particle. Some examples of this category are as the following.

$$_3\text{Li}^6 + {_0}\text{n}^1 \rightarrow (_3\text{Li}^7)^* \rightarrow {_1}\text{H}^3 + {_2}\text{He}^4$$

$$_5\text{B}^{10} + {_0}\text{n}^1 \rightarrow (_5\text{B}^{11})^* \rightarrow {_3}\text{Li}^7 + {_2}\text{He}^4$$

$$_{13}\text{Al}^{27} + {_0}\text{n}^1 \rightarrow (_{13}\text{Al}^{28})^* \rightarrow {_{11}}\text{Na}^{24} + {_2}\text{He}^4$$

$$_7\text{N}^{14} + {_0}\text{n}^1 \rightarrow (_7\text{N}^{15})^* \begin{array}{l} \nearrow\ {_5}\text{B}^{11} + {_2}\text{He}^4 \\ \searrow\ {_3}\text{Li}^7 + {_2}\text{He}^4 + {_2}\text{He}^4 \end{array}$$

When fast neutrons are captured by a heavy nucleus, an alpha particle is emitted and the product nucleus is usually radioactive. Some examples are as the following.

$$_{11}\text{Na}^{23} + {_0}\text{n}^1 \rightarrow (_{11}\text{Na}^{24})^* \rightarrow {_9}\text{F}^{20} + {_2}\text{He}^4$$

Here, $_9\text{F}^{20}$ is a radioactive element which decays in the following manner.

$$_9\text{F}^{20} \rightarrow {_{10}}\text{Ne}^{20} + \beta^-$$

Another example is

$$_{13}\text{Al}^{27} + {_0}\text{n}^1 \rightarrow (_{13}\text{Al}^{28})^* \rightarrow {_{11}}\text{Na}^{24} + {_2}\text{He}^4$$

Here, $_{11}\text{Na}^{24}$ is a radioactive element which decays in the following manner.

$$_{11}\text{Na}^{24} \rightarrow {_{12}}\text{Mg}^{24} + \beta^-$$

(b) (n, p) reaction: In such reactions, the outgoing particle is a proton. Thus, in the nucleus, a proton is replaced by a neutron. Here, the Q value is positive. Moreover, the product nucleus decays further. Some examples of this category are as the following.

$$_7\text{N}^{14} + {_0}\text{n}^1 \rightarrow (_7\text{N}^{15})^* \rightarrow {_6}\text{C}^{14} + {_1}\text{H}^1$$

The product nucleus $_6\text{C}^{14}$ decays in the following manner.

$$_6\text{C}^{14} \rightarrow {_7}\text{N}^{14} + \beta^-$$

Another example is

$$_2He^3 + {_0n^1} \rightarrow ({_2He^4})^* \rightarrow {_1H^3} + {_1H^1}$$

The product nucleus $_1H^3$ decays in the following manner.

$$_1H^3 \rightarrow {_2He^3} + \beta^-$$

Further, example is

$$_{17}Cl^{35} + {_0n^1} \rightarrow ({_{17}Cl^{36}})^* \rightarrow {_{16}S^{35}} + {_1H^1}$$

The product nucleus $_{16}S^{35}$ decays in the following manner.

$$_{16}S^{35} \rightarrow {_{17}Cl^{35}} + \beta^-$$

When the Q value is negative, the fast neutrons produce the following (n, p) reactions. Some examples of such reactions are as the following.

$$_{13}Al^{27} + {_0n^1} \rightarrow ({_{13}Al^{28}})^* \rightarrow {_{12}Mg^{27}} + {_1H^1}$$

$$_{16}S^{32} + {_0n^1} \rightarrow ({_{16}S^{33}})^* \rightarrow {_{15}P^{32}} + {_1H^1}$$

$$_{29}Cu^{65} + {_0n^1} \rightarrow ({_{29}Cu^{66}})^* \rightarrow {_{28}Ni^{65}} + {_1H^1}$$

$$_{30}Zn^{64} + {_0n^1} \rightarrow ({_{30}Zn^{65}})^* \rightarrow {_{29}Cu^{64}} + {_1H^1}$$

For all these reactions, the product nucleus is radioactive, which decays finally to the initial target nucleus.

(c) Radiative capture: In such reactions, the outgoing particle is a γ-radiation. Some examples of this category are as the following.

$$_1H^1 + {_0n^1} \rightarrow ({_1H^2})^* \rightarrow {_1H^2} + \gamma$$

$$_1H^2 + {_0n^1} \rightarrow ({_1H^3})^* \rightarrow {_1H^3} + \gamma$$

With the capture of a neutron, the neutron to proton ratio increases. When this ratio in the target nucleus is near the upper limit of stability, the product nucleus obviously becomes radioactive and generally emits a β-particle. Some examples of this category are as the following.

$$_{45}Rh^{103} + {_0n^1} \rightarrow ({_{45}Rh^{104}})^* \rightarrow {_{45}Rh^{104}} + \gamma$$

$$_{49}In^{115} + {_0n^1} \rightarrow ({_{49}In^{116}})^* \rightarrow {_{49}In^{116}} + \gamma$$

$$_{92}U^{238} + {_0n^1} \rightarrow ({_{92}U^{239}})^* \rightarrow {_{92}U^{239}} + \gamma$$

$$_{29}\text{Cu}^{65} + {_0}\text{n}^1 \rightarrow ({_{29}}\text{Cu}^{66})^* \rightarrow {_{29}}\text{Cu}^{66} + \gamma$$

$$_{79}\text{Au}^{197} + {_0}\text{n}^1 \rightarrow ({_{79}}\text{Au}^{198})^* \rightarrow {_{79}}\text{Au}^{198} + \gamma$$

Here, each of the product nuclei is radioactive.

(d) (n,d) and (n,t) reactions: (n,d) reactions are obtained as a result of the bombardment of elements by 90 MeV neutrons. In these reactions, compound nucleus is not produced and thus, the reactions are of pick-up type. The (n,t) reactions can also take place in a similar manner. One of the examples of such reactions is as the following.

$$_7\text{N}^{14} + {_0}\text{n}^1 \rightarrow {_6}\text{C}^{12} + {_1}\text{H}^3$$

(e) $(n,2n)$ reactions: In such type of reaction, one neutron is captured by the target nucleus and two neutrons are emitted, so that the charge of the product nucleus is the same as that of the target nucleus. However, the mass reduces by one. Thus, the product nucleus is the isotope of the target nucleus. Q-value of the reaction is negative and therefore fast neutrons are required for the reaction. In most of the cases, the residual nucleus is unstable which later on emits a positron. Some examples of such reactions are as the following.

$$_6\text{C}^{12} + {_0}\text{n}^1 \rightarrow ({_6}\text{C}^{13})^* \rightarrow {_6}\text{C}^{11} + 2\,{_0}\text{n}^1$$

$$_{19}\text{K}^{39} + {_0}\text{n}^1 \rightarrow ({_{19}}\text{K}^{40})^* \rightarrow {_{19}}\text{K}^{38} + 2\,{_0}\text{n}^1$$

$$_{51}\text{Sb}^{121} + {_0}\text{n}^1 \rightarrow ({_{51}}\text{Sb}^{122})^* \rightarrow {_{51}}\text{Sb}^{120} + 2\,{_0}\text{n}^1$$

(f) (n, three or more particles) reaction: When the energy of the incident neutron approaches to 30 MeV, three neutrons or even two neutrons and a proton are ejected from the compound nucleus.

8.4 Transmutation by deuteron

In such transmutation a deuteron is absorbed, forming a compound nucleus. In the process of disintegration of this compound nucleus, the outgoing particle may be an alpha particle, proton, neutron, gamma radiation or other particle.

(a) (d,α) reaction: In such reactions, the outgoing particle is an alpha particle. Some examples of this category are as the following.

$$_3\text{Li}^6 + {_1}\text{H}^2 \rightarrow ({_4}\text{Be}^8)^* \rightarrow {_2}\text{He}^4 + {_2}\text{He}^4$$

$$_8\text{O}^{16} + {_1}\text{H}^2 \rightarrow ({_9}\text{F}^{18})^* \rightarrow {_7}\text{N}^{14} + {_2}\text{He}^4$$

$$_{10}Ne^{20} + {_1H^2} \rightarrow ({_{11}Na^{22}})^* \rightarrow {_9F^{18}} + {_2He^4}$$

$$_{12}Mg^{26} + {_1H^2} \rightarrow ({_{13}Al^{28}})^* \rightarrow {_{11}Na^{24}} + {_2He^4}$$

$$_{13}Al^{27} + {_1H^2} \rightarrow ({_{14}Si^{29}})^* \rightarrow {_{12}Mg^{25}} + {_2He^4}$$

(b) (d, p) reaction: In such reactions, the outgoing particle is a proton. Some examples of this category are as the following.

$$_6C^{12} + {_1H^2} \rightarrow ({_7N^{14}})^* \rightarrow {_6C^{13}} + {_1H^1}$$

$$_3Li^7 + {_1H^2} \rightarrow ({_2Be^9})^* \rightarrow {_3Li^8} + {_1H^1}$$

$$_{11}Na^{23} + {_1H^2} \rightarrow ({_{12}Mg^{25}})^* \rightarrow {_{11}Na^{24}} + {_1H^1}$$

$$_{15}P^{31} + {_1H^2} \rightarrow ({_{16}S^{33}})^* \rightarrow {_{15}P^{32}} + {_1H^1}$$

$$_{48}Cd^{114} + {_1H^2} \rightarrow ({_{49}In^{116}})^* \rightarrow {_{48}Cd^{115}} + {_1H^1}$$

$$_{83}Bi^{209} + {_1H^2} \rightarrow ({_{84}Po^{211}})^* \rightarrow {_{83}Bi^{210}} + {_1H^1}$$

(c) (d, n) reaction: In such reactions, the outgoing particle is a neutron. Some examples of this category are as the following.

$$_3Li^7 + {_1H^2} \rightarrow ({_4Be^9})^* \rightarrow {_4Be^8} + {_0n^1}$$

$$_4Be^9 + {_1H^2} \rightarrow ({_5B^{11}})^* \rightarrow {_5B^{10}} + {_0n^1}$$

$$_6C^{12} + {_1H^2} \rightarrow ({_7N^{14}})^* \rightarrow {_7N^{13}} + {_0n^1}$$

When two deuterons interact, both (d, n) and (d, p) reactions have been observed:

$$_1H^2 + {_1H^2} \rightarrow ({_2He^4})^* \begin{array}{l} \nearrow \; {_1H^3} + {_1H^1} \\ \searrow \; {_2He^3} + {_0n^1} \end{array}$$

(d) (d, t) reaction: Though the probability of such reaction is low, some examples of them are however known. Some of them are as the following.

$$_3Li^7 + {_1H^2} \rightarrow ({_4Be^9})^* \rightarrow {_3Li^6} + {_1H^3}$$

$$_{15}P^{31} + {_1H^2} \rightarrow ({_{16}S^{33}})^* \rightarrow {_{15}P^{30}} + {_1H^3}$$

$$_{47}Ag^{109} + {_1H^2} \rightarrow ({_{48}Cd^{111}})^* \rightarrow {_{47}Ag^{108}} + {_1H^3}$$

$$_4Be^9 + {_1H^2} \rightarrow ({_5B^{11}})^* \rightarrow {_4Be^8} + {_1H^3}$$

8.5 Transmutation by particles of mass three

Some reactions where tritium ($_1H^3$) or helium 3 ($_2He^3$) are used as projectile are as the following.

$$_3Li^6 + {_1H^3} \rightarrow (_4Be^9)^* \rightarrow {_3Li^7} + {_1H^2}$$

$$_{27}Co^{59} + {_1H^3} \rightarrow (_{28}Ni^{62})^* \rightarrow {_{27}Co^{60}} + {_1H^2}$$

$$_{29}Cu^{63} + {_1H^3} \rightarrow (_{30}Zn^{66})^* \rightarrow {_{29}Cu^{64}} + {_1H^2}$$

$$_1H^2 + {_1H^3} \rightarrow (_2He^5)^* \rightarrow {_2He^4} + {_0n^1}$$

$$_{16}S^{32} + {_1H^3} \rightarrow (_{17}Cl^{35})^* \rightarrow {_{17}Cl^{34}} + {_0n^1}$$

$$_{13}Al^{27} + {_1H^3} \rightarrow (_{14}Sl^{30})^* \rightarrow {_{12}Mg^{27}} + {_2He^3}$$

$$_3Li^7 + {_1H^3} \rightarrow (_4Be^{10})^* \rightarrow {_2He^6} + {_2He^4}$$

$$_1H^2 + {_2He^3} \rightarrow (_3Li^5)^* \rightarrow {_2He^4} + {_1H^1}$$

$$_1H^3 + {_2He^3} \rightarrow (_3Li^6)^* \rightarrow {_2He^5} + {_1H^1}$$

$$_2He^3 + {_2He^3} \rightarrow (_4Be^6)^* \rightarrow {_3Li^5} + {_1H^1}$$

$$_7N^{14} + {_2He^3} \rightarrow (_9F^{17})^* \rightarrow {_9F^{16}} + {_0n^1}$$

$$_6C^{12} + {_2He^3} \rightarrow (_8O^{15})^* \rightarrow {_7N^{13}} + {_1H^2}$$

$$_6C^{13} + {_2He^3} \rightarrow (_8O^{16})^* \rightarrow {_7N^{13}} + {_1H^3}$$

$$_7N^{14} + {_2He^3} \rightarrow (_9F^{17})^* \rightarrow {_7N^{13}} + {_2He^4}$$

8.6 Transmutation by Radiation

Here, γ-radiation behave as an incident particle. For this process, the energy of the incident photon must be more than the binding energy of the particle to be ejected from the target nucleus. When the energy of the γ-radiation is not sufficient to remove even neutron, the nucleus may enter in one of its excited states. The excess energy is then emitted as a γ-radiation. Some examples of such reactions are as the following.

$$_1H^2 + \gamma \rightarrow (_1H^2)^* \rightarrow {_1H^1} + {_0n^1}$$

$$_4Be^9 + \gamma \rightarrow (_4Be^9)^* \rightarrow {_4Be^8} + {_0n^1}$$

$$_{15}P^{31} + \gamma \rightarrow (_{15}P^{31})^* \rightarrow {_{15}P^{30}} + {_0n^1}$$

$$_4Be^9 + \gamma \rightarrow (_4Be^9)^* \rightarrow {_3Li^8} + {_1H^1}$$

$$_{12}Mg^{25} + \gamma \rightarrow (_{12}Mg^{25})^* \rightarrow {_{11}Na^{24}} + {_1H^1}$$

$$_5B^{10} + \gamma \rightarrow (_5B^{10})^* \rightarrow {_4Be^8} + {_1H^2}$$

$$_5B^{11} + \gamma \rightarrow (_5B^{11})^* \rightarrow {_4Be^8} + {_1H^3}$$

$$_6C^{12} + \gamma \rightarrow (_6C^{12})^* \rightarrow {_4Be^8} + {_2He^4}$$

$$_{36}Kr^{83} + \gamma \rightarrow (_{36}Kr^{83})^* \rightarrow {_{36}Kr^{83}} + \gamma'$$

$$_{38}Sr^{87} + \gamma \rightarrow (_{38}Sr^{87})^* \rightarrow {_{38}Sr^{87}} + \gamma'$$

When energy of γ-radiation is of the order of 10 MeV, one nucleon, either neutron or proton can be emitted. For higher energies, we can have reactions of the types (γ, np), $(\gamma, 2n)$, $(\gamma, n2p)$ and others, where two or more nucleons are emitted. Some examples of such type of reaction are as the following.

$$_8O^{16} + \gamma \rightarrow (_8O^{16})^* \rightarrow {_6C^{11}} + 3\ {_0n^1} + 2\ {_1H^1}$$

$$_6C^{12} + \gamma \rightarrow (_6C^{12})^* \rightarrow 3\ {_2He^4}$$

Exercise 3. In a nuclear reactor, deuterium undergoes the reaction $_1H^2$ $(d,\ p)$ $_1H^3$ followed by the reaction $_1H^3$ $(d,\ n)$ $_2He^4$. (i) Calculate the amount of energy released in each of these reactions. (ii) Calculate the total energy released per gram of deuteron used in fusion. Given, mass of $_1H^1$ = 1.007825 amu, mass of $_1H^2$ = 2.014102 amu, mass of $_1H^3$ = 3.016049 amu, mass of $_0n^1$ = 1.008665 amu, mass of $_2He^4$ = 4.002604 amu.

Solution: (i) For the reaction $_1H^2$ $(d,\ p)$ $_1H^3$, we have

$$_1H^2 + {_1H^2} \rightarrow {_1H^3} +_1 H^1 + Q$$

and the mass defect is Δm = 2.014102 + 2.014102 - 3.016049 - 1.007825 = 0.00433 amu

Thus, the energy released is E = 0.00433 × 931.3 = 4.033 MeV

For the reaction $_1H^3$ $(d,\ n)$ $_2He^4$, we have

$$_1H^3 + {_1H^2} \rightarrow {_0n^1} +_2 He^4 + Q$$

and the mass defect is $\Delta m = 3.016049 + 2.014102 - 4.002604 - 1.008665 = 3.016049 + 2.014102 - 4.002604 - 1.008665 = 0.018882$ amu

Thus, the energy released is $E = 0.018882 \times 931.3 = 17.585$ MeV

(ii) Total energy released in these two successive reactions is $= 4.033 + 17.585 = 21.618$ MeV

This total energy is released due to fusion of 3 deuteron. Hence, the energy released per deuteron is $= 21.618/3 = 7.206$ MeV

The number of nuclei in 1 gm of deuteron is $= 6.025 \times 10^{23}/2 = 3.0125 \times 10^{23}$

Thus, total energy released from 1 gm of deuteron is $= 3.0125 \times 10^{23} \times 7.206 = 2.17 \times 10^{24}$ MeV

Exercise 4. For the reaction

$$H^1 + F^{19} \rightarrow He^4 + O^{16}$$

masses of all nuclei are determined accurately with the help of mass spectrometer as: $m_H = 1.007825$ amu, $m_{He} = 4.002603$ amu, $m_F = 18.998405$ amu, $m_O = 15.994915$ amu. Calculate the Q-value and nature of the reaction.

Solution: For the reaction

$$H^1 + F^{19} \rightarrow He^4 + O^{16} + Q$$

We have

$$Q = 1.007825 + 18.998405 - 4.002603 - 15.994915 = 0.008712 \text{ amu}$$

Therefore, the energy released is $Q = 0.008712 \times 931 = 8.11$ MeV. As the energy released is positive, the reaction is exothermic.

Exercise 5. For the reaction

$$N^{14} + He^4 \rightarrow O^{17} + H^1$$

masses of all nuclei are determined accurately with the help of mass spectrometer as: $m_N = 14.003074$ amu, $m_{He} = 4.002603$ amu, $m_O = 16.99913$ amu, $m_H = 1.007825$ amu. Calculate the Q-value and nature of the reaction.

Solution: For the reaction

$$N^{14} + He^4 \rightarrow O^{17} + H^1 + Q$$

We have

$$Q = 14.003074 + 4.002603 - 16.99913 - 1.007825 = -0.001277 \text{ amu}$$

Therefore, the energy released is $Q = -0.001277 \times 931 = -\ 1.189$ MeV. As the energy released is negative, the reaction is endothermic.

Exercise 5. Calculate the Q-value for the reactions Al^{27} (d, *alpha*) Mg^{25} and Mg^{25} (*alpha*, d) Al^{27}. Given, masses of all nuclei are determined accurately with the help of mass spectrometer as: $m_{Al} = 26.9901$ amu, $m_{Mg} = 24.9936$ amu, $m_\alpha = 4.0039$ amu, $m_d = 2.0147$ amu.

Solution: For the reaction

$$Al^{27} + H^2 \rightarrow Mg^{25} + He^4 + Q$$

We have

$$Q = 24.9936 + 4.0039 - 26.9901 - 2.0147 = -0.0073 \text{ amu}$$

Therefore, the energy released is $Q = -0.0073 \times 931 = -\ 6.796$ MeV. As the energy released is negative, the reaction is endothermic.

For the reaction

$$Mg^{25} + He^4 \rightarrow Al^{27} + H^2 + Q$$

We have

$$Q = 26.9901 + 2.0147 - 24.9936 - 4.0039 = 0.0073 \text{ amu}$$

Therefore, the energy released is $Q = 0.0073 \times 931 = 6.796$ MeV. As the energy released is positive, the reaction is exothermic.

9. Problems and questions

1. Derive an expression for Q-value of a nuclear reaction. Show that an exothermic reaction is possible even when the reacting nuclei have no energy.

2. Derive an expression for the Q-equation and show that an endothermic reaction requires a minimum amount of energy of the incident particle.

3. Discuss about a compound nucleus. Describe about the virtual levels in a compound nucleus.

4. Discuss about the cross sections for nuclear reactions. State the salient features of Breit-Wigner formula. With the help of the Breit-Wigner formula show that the full width at half-maximum is equal to the total width.

5. Write notes on the following

   (i) Elastic scattering and inelastic scattering
   (ii) Pick-up ans stripping reactions
   (iii) Disintegration
   (iv) Photo disintegration
   (v) Radiative capture
   (vi) Spontaneous decay
   (vii) Spallation reactions
   (viii) High energy reactions
   (ix) Exothermic reaction
   (x) Endothermic reaction
   (xi) Compound nucleus model for nuclear reaction
   (xii) Breit-Wigner formula
   (xiii) Optical Model for nuclear reaction

V. Fission and Fusion

After the discovery of neutron in 1932 by Chadwick, a systematic investigation of nuclear reactions induced by neutrons was started by E. Fermi and his collaborators. During such investigation, Otto Hahn and Fritz Strassmann noticed that when the uranium was bombarded by neutrons, the uranium nucleus broken up into two lighter nuclei. Lise Meitner and Otta Frisch successfully explained the results of these experiments with the help of liquid drop model for a nucleus, proposed by Niels Bohr. They called this process of breaking nucleus as the 'Nuclear Fission'. In the fission process, a heavy nucleus breaks up into two light nuclei. On the other side, there is a nuclear process where some light nuclei fuse together to form a heavy nucleus. This process is known as the 'nuclear fusion'. In this chapter, we shall discuss about these two nuclear phenomena.

1. Nuclear fission

When a uranium nucleus is bombarded by a neutron, the uranium nucleus splits into two light nuclei. In case, one of them is barium (Z = 56), the other one must be krypton (Z = 36). According to the compound nucleus theory, the neutron is captured by the uranium and a compound nucleus is formed. This compound nucleus then breaks in two parts. The reaction may be expressed as (Figure 1)[1]

$$ {}_{92}\mathrm{U}^{235} + {}_{0}\mathrm{n}^{1} \rightarrow ({}_{92}\mathrm{U}^{236})^{*} \rightarrow {}_{56}\mathrm{Ba}^{141} + {}_{36}\mathrm{Kr}^{92} + 3\,{}_{0}\mathrm{n}^{1} + Q $$

This type of disintegration process in which a heavy nucleus splits into two nuclei of comparable masses is known as the nuclear fission. Here,

[1]The compounds nucleus may also break into molybdenum and xenon as

$$ {}_{92}\mathrm{U}^{235} + {}_{0}\mathrm{n}^{1} \rightarrow ({}_{92}\mathrm{U}^{236})^{*} \rightarrow {}_{42}\mathrm{Mo}^{98} + {}_{54}\mathrm{Xe}^{136} + 2\,{}_{0}\mathrm{n}^{1} + Q $$

The charge is conserved through the emission of four electrons.

Q denotes the amount of energy released in the fission process. The asterisk indicates the compound nucleus in the excited state.

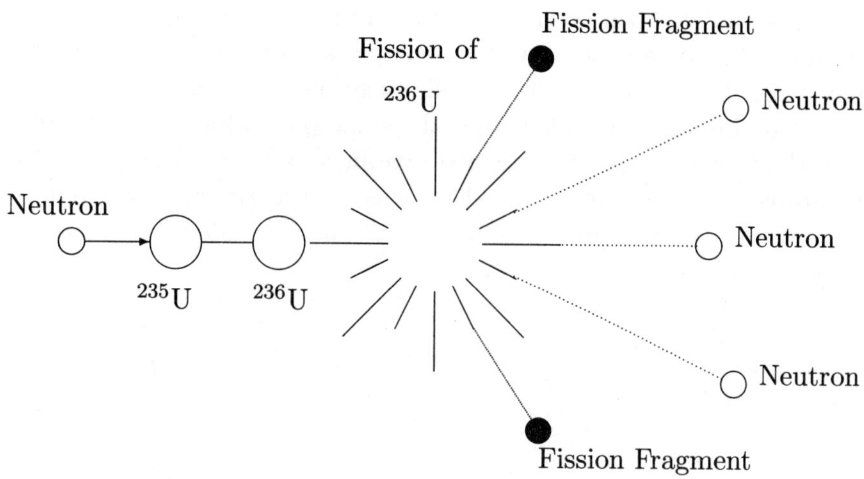

Figure 1: Schematic representation of nuclear fission of U^{235} be capturing a neutron.

1.1 Bohr and Wheeler theory for nuclear fission

Soon after the discovery of disintegration of uranium nucleus, bombarded by a neutron, into two light nuclei, the explanation for the process was given with the help of the liquid drop model suggested by Niels Bohr. In the liquid drop model, the shape of the drop depends on a balance between the surface tension of the liquid drop and the Coulomb repulsion among the protons. (The surface tension tries to bring the nucleus so that its surface area is as minimum as possible. The Coulomb repulsion among the protons tries to increase the volume of the drop.) In the normal state, the shape of a nucleus is spherical. When the energy is added to the nucleus, resulting from the capture of a slow neutron, the energy appears in the form of excitation energy. As a result of that the drop is deformed and due to the two opposite forces working simultaneously, the oscillations are set in the drop. The shape of the drop is distorted from the spherical one and may become ellipsoidal. The force due to the surface tension, tries to return the drop to its original shape (spherical shape) while the excitation energy tends to distort as much as possible.

Now, there are two possibilities: (1) When the excitation energy is not sufficiently large and the surface tension force is more strong, then the nucleus will never reach to the shape of dumb-bell (stage (iii) in

Figure 2). It will make oscillations and each time will loose energy of excitation. After some oscillations, the nucleus will finally return back to its original spherical shape. The energy is generally released in the form of γ-rays and this will be the process of radiative capture of neutron by uranium. (2) When the excitation energy is sufficiently large, the drop may attain the shape of a dumb-bell (stage (iii) in Figure 2). In case the oscillation becomes so violent that the stage (iv) in Figure 2 is attained, then the fission of the nucleus into two nuclei (stage (v) in Figure 2) is inevitable. Thus, there is a threshold energy or a critical energy required to produce the stage (iv), after which the nucleus cannot return back to the stage (i).

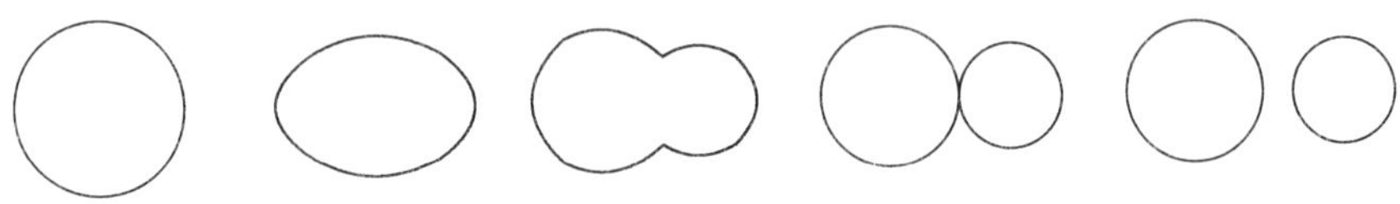

Figure 2: Schematic representation of nuclear fission process based on the liquid drop model.

Schematic representation of nuclear fission based on the liquid drop model is shown in Figure 2. This is also known as the Bohr and Wheeler theory for nuclear fission. For the spherical shape, we have only one center. Due to deformation, the drop assumes the shape of ellipsoid and there are two distinct centers in the drop. With the increase of deformation, the separation between these centers increases and in the state of fission process, they become centers of two independent fragments, which finally assume the spherical shape (stage (v) in Figure 2).

Potential energy of a drop in different stages (shapes) can be calculated as a function of the degree of deformation (*i.e.*, the separation between the two centers) of the drop. The potential energy, as a function of the separation r between the centers of two fragments is shown in Figure 3 for three nuclei of different mass numbers A. $A = 236$ is the smallest nucleus showing the process of nuclear fission. We have considered one nucleus with $A < 236$ and another nucleus with $A > 236$. In the figure, Q is the threshold energy required for fission of the nucleus into two fragments. The stage (iv) in Figure 2 is shown by the vertical dotted line. It is denoted by D there.

A heavy nucleus ($A \geq 300$) breaks up into fragments instantaneously as soon as it is formed. The reason for it is the large amount of coulomb forces acting among the protons in the nucleus. This is shown by the graph (a) in Figure 3. For $A \approx 236$, variation of potential energy is shown by the graph (b) in Figure 3. It shows that for the separation smaller than a certain value marked by D (vertical line), the strong attractive nuclear forces dominate the repulsive Coulomb forces and the potential energy follows the curve BC. In this region, the oscillations are stable and the nucleus can return back to its original spherical shape. For the deformation larger than D, the Coulomb forces dominate and with the increase in the deformation, the separation between the charge centers becomes more, as a result the potential energy decreases. It is shown by the curve CK. At the critical deformation D, the potential energy is known as a fission barrier or critical energy for the fission. It corresponds to the breaking of the parent nucleus into two fragments, called the fission products.

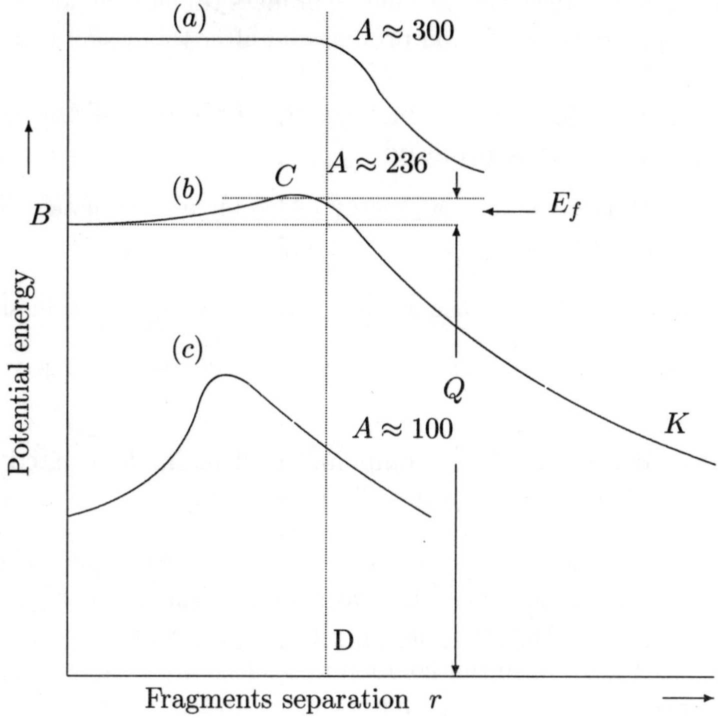

Figure 3: Variation of potential energy as a function of separation r between the centers of two segments for three nuclei with different mass numbers A.

The critical energy is the difference the difference between the energy of nucleus at point C and the initial energy of nucleus at B. It is represented by E_f. When the atomic mass number of a nucleus is further low, for example, $A \approx 100$, the fission barrier is much higher, as shown in the graph (c). In this case, when a neutron is captured, it does not give sufficient excitation energy to surmount the barrier.

It is interesting to note that classically, the barrier hinders the nucleus to split when the deformation lies below the region BC. However, we know that quantum mechanically, there is always a finite probability for a particle to tunnel through a potential barrier in such cases. It shows that even when the energy of deformation is below E_f, there is finite probability for the fission process to take place, though it will occur very slowly. This is the situation of spontaneous fission of a nucleus, which otherwise is stable against the fission. (In the natural uranium, the nucleus ^{238}U undergoes spontaneous fission with a half-life of about 10^{16} years.)

In the case of the fission process, the two fragments suffer Coulomb repulsion and at a finite separation between them, their kinetic energy can be estimated in terms of the change in mass (difference between the mass of bodies before the fission process and after the fission process).

1.2 Calculation of critical energy on the basis of the liquid drop model of a nucleus

In order to estimate the critical energy of fission E_f, we make the following assumptions:

1. The initial nucleus is in the ground state and is spherical in shape.
2. The total volume of the system remains constant during various stages
3. Surface oscillations of the compound nucleus in the excited state produce deformation in the surface.

As a simplification of the treatment, let us consider the process of fission as symmetric one, *i.e.*, the two fission fragments to be alike in charge and mass, which is not the case in actual practice. Hence, when the charge and mass of initial compound nucleus are Z and A, respectively, then each fragment will have charge and mass $Z/2$ and $A/2$, respectively. (We assume both Z and A to be even numbers, so that they could be divided by 2.) Let us first calculate total energy of the initial drop, as the following.

(i) Surface energy of the initial spherical drop of radius R is

$$E_{si} = 4\pi R^2 T$$

where T is the surface tension.

(ii) The electrostatic energy of the initial drop is

$$E_{ei} = \frac{3}{5}\frac{(Ze)^2}{R}$$

Thus, the total energy is the sum of E_{si} and E_{ei} and we have

$$E_i = 4\pi R^2 T + \frac{3}{5}\frac{(Ze)^2}{R}$$

Using the relation $R = R_0 A^{1/3}$, where $R_0 = 1.2 \times 10^{-15}$ m, we get

$$E_i = 4\pi R_0^2 A^{2/3} T + \frac{3}{5}\frac{(Ze)^2}{R_0 A^{1/3}} \tag{5.1}$$

After the fission process, the radius of each fragment is $r = R_0(A/2)^{1/3}$. Let us now calculate total energy of the fragments after the fission process, as the following.

(iii) Surface energy of both the fragments, each of radius r is

$$E_{sp} = 2 \times 4\pi r^2 T = 8\pi R_0^2 (A/2)^{2/3} T$$

(iv) Now, the electrostatic energy has two components: (a) energy of formation of each component separately and (b) the interaction energy between the two components. Thus, we have

$$E_{ep} = 2 \times \frac{3}{5}\frac{(Ze/2)^2}{r} + \frac{(Ze/2)^2}{2r} = \frac{3}{10}\frac{Z^2e^2}{R_0(A/2)^{1/3}} + \frac{Z^2e^2}{8R_0(A/2)^{1/3}}$$

Thus, the total energy is the sum of E_{sp} and E_{ep} and we have

$$E_p = 8\pi R_0^2 (A/2)^{2/3} T + \frac{3}{10}\frac{Z^2e^2}{R_0(A/2)^{1/3}} + \frac{Z^2e^2}{8R_0(A/2)^{1/3}}$$

The fission energy or the critical energy E_f for fission is the difference between E_p and E_i. Thus, we have

$$E_f = E_p - E_i = 8\pi R_0^2 (A/2)^{2/3} T + \frac{3}{10}\frac{Z^2e^2}{R_0(A/2)^{1/3}} + \frac{Z^2e^2}{8R_0(A/2)^{1/3}}$$

$$-4\pi R_0^2 A^{2/3} T - \frac{3}{5}\frac{(Ze)^2}{R_0 A^{1/3}} = 4\pi R_0^2 A^{2/3} T a - \frac{3}{10}\frac{Z^2e^2}{R_0 A^{1/3}} b \tag{5.2}$$

where the constants $a = (2^{1/3} - 1) = 0.26$ and $b = (2 - 2^{1/3} \times 1.4166) = 0.22$ are constants. In the first approximation, a and b can be taken to be equal.

For the spontaneous fission to take place, we should have $E_f = 0$. Thus, from equation (5.2), we have

$$4\pi R_0^2 A^{2/3} Ta - \frac{3}{10}\frac{Z^2 e^2}{R_0 A^{1/3}} b = 0$$

It gives

$$\frac{Z^2}{A} = \frac{2 \times 4\pi R_0^2 T}{3e^2/5R_0}$$

From the Weiszacker's mass formula, we know $4\pi R_0^2 T = 17.8$ and $3e^2/5R_0 = 0.71$. On using these values, we get

$$\frac{Z^2}{A} = \frac{2 \times 17.8}{0.71} \approx 50$$

However, a more general acceptable value for this parameter Z^2/A for spontaneous fission is 47.8. Thus, a nucleus will undergo the fission instantaneously when Z^2/A is close to this value. When Z^2/A is smaller than this value, the nucleus will be stable against the spontaneous fission. For U^{235}, the value of Z^2/A is 36.02 and for Pu^{239}, it is 36.97, showing that that U^{235} as well as Pu^{239} cannot undergo spontaneous fission. For the fission process of these nuclei, some required amount of energy need to be supplied through the capture of a projectile. For some nuclei, critical energy is given in Table 1.

Table 1. Critical energy for some nuclei

Target	Z^2/A	E_f (MeV)	
		Calculated	Observed
$_{90}Th^{230}$	34.91	6.5	5.95
$_{92}U^{233}$	36.33	4.6	5.49
$_{92}U^{235}$	36.02	5.3	5.75
$_{90}U^{238}$	35.56	5.5	5.80
$_{94}Pu^{239}$	36.97	4.0	5.48

1.3 Fission products and energy released

Let us consider the fission of $_{92}U^{235}$ through the capture of a thermal neutron as the following.

$$_{92}U^{235} + {}_0n^1 \rightarrow ({}_{92}U^{236})^* \rightarrow 2 \text{ Fission fragments} + x \text{ neutrons} + Q$$

Here, the asterisk denotes the compound nucleus. An amount of energy Q is released and x is the number of neutrons produced. For the conservation of charge, β particles (electrons or positrons) are also produced.

The average value of x, generally, lies between 2 and 3, depending on the fission fragments. The fission fragments are generally of unequal masses, one of them is heavier than the other. Such a fission is known as the asymmetric fission.

For the fission of a nucleus, the fragments may be different, depending on the channel followed by the process. For example, in case of the fission of ${}_{92}\mathrm{U}^{235}$, there are 40 different ways, producing different nuclei. Out of them, two examples are as the following.

$$
\begin{aligned}
{}_{92}\mathrm{U}^{235} + {}_0\mathrm{n}^1 \rightarrow ({}_{92}\mathrm{U}^{236})^* &\rightarrow {}_{56}\mathrm{Ba}^{141} + {}_{36}\mathrm{Kr}^{92} + 3\,{}_0\mathrm{n}^1 + Q \\
&\rightarrow {}_{54}\mathrm{Xe}^{140} + {}_{38}\mathrm{Sr}^{94} + 2\,{}_0\mathrm{n}^1 + Q
\end{aligned}
$$

The mass number A of the heavier fragment generally lies in the range from 125 to 150, whereas of the lighter fragment lies in the range from 80 to 100. The mass distribution of fission fragments is mostly determined in the form of yield. In the lighter group, nuclei of mass number A around 95 has maximum yield 6.3% and the corresponding mass A around 140 has maximum yield 6.3%. These peaks around $A = 95$ and $A = 140$ are associated with neutron number around $N = 50$ and $N = 82$. Thus, heavy nuclei have a tendency to break into fragments having closed shells with magic numbers 50 and 82, respectively.

Since the linear momentum is conserved, the fission fragments move exactly in opposite directions, carrying most of the momentum such that $m_1v_1 = m_2v_2$. Here, m_1 and m_2 are masses, and v_1 and v_2 of two fragments, respectively. Here, we have used the fact that the compound nucleus is stationary. Thus, the ratio of the average kinetic energies of the fission fragments is

$$\frac{E_2}{E_1} = \frac{m_2v_2^2/2}{m_1v_1^2/2} = \frac{m_1}{m_2}$$

It shows that the distribution of kinetic energy between the fission fragments is inversely proportional to their masses. That is the heavier fragment carries less energy than the lighter fragment. Moreover, the heavier one moves with lower velocity as compared to the lighter one.

It has been found that symmetric fission due to thermal neutrons is very rare. However, it becomes probable with the increase of the energy of incident neutron.

The energy released during fission process is distributed as kinetic energy of fission fragments, prompt neutrons, γ-rays both prompt and delayed, β-particles (electrons or positrons) and neutrinos. On average, total energy released in one event of fission lies in the range from 190 to 210 MeV. However, for all practical purposes, we take it 200 MeV per fission. The energy released in the fission is dissipated into the surrounding environment in the form of heat.

For an estimation of energy, let us consider an example as the following reaction.

$$_{92}U^{235} + {_0n^1} \rightarrow {_{92}U^{236}} \rightarrow {_{42}Mo^{98}} + {_{54}Xe^{136}} + 2\,{_0n^1}$$

The charge is conserved through the emission of four β particles (electrons). Mass of a β particle may be taken as negligible in comparison to other masses. On the left side of equation, we have

$$\begin{aligned} {_{92}U^{235}} &= 235.044 \\ {_0n^1} &= \underline{\quad 1.009} \\ & \quad \underline{236.053} \end{aligned}$$

The total mass before the fission reaction is 236.053 amu. Now, on the right side of equation, we have

$$\begin{aligned} {_{42}Mo^{98}} &= 97.906 \\ {_{54}Xe^{136}} &= 135.907 \\ 2\,{_0n^1} &= \underline{\quad 2.018} \\ & \quad \underline{235.831} \end{aligned}$$

The total mass after the fission reaction is 235.840 amu.

Thus, the mass defect = 236.053 – 235.831 = 0.222 amu

Hence, the energy released = 0.222 × 931.3 = 206.75 MeV

This value for the released energy is typical of all fission energies, so that we generally use the approximate value of 200 MeV per fission. This energy for each fission is distributed roughly as the following.

K.E. of fission fragments	168	MeV
K.E. of fast fission neutrons	5	MeV
β-decay energy from fission products	4.8	MeV
Neutrino energy from β-decays	10.0	MeV
Intermediate γ-rays energy	4.6	MeV
Fission product γ-rays energy	6.9	MeV
Total	199.3	MeV

Since 1 MeV is equivalent to 1.602 $\times 10^{-13}$ J, the energy released per fission of U^{235} is 3.204 $\times 10^{-11}$ J, which is very small. There are 6.025 $\times 10^{23}$ atoms in 235 gm U^{235}. Assuming that all the nuclei of these

atoms undergo fission, the energy E produced per gram of U^{235} is

$$E = \frac{3.204 \times 10^{-11} \times 6.025 \times 10^{23}}{235} = 8.21 \times 10^{10}\text{J}$$

Exercise 1: A nuclear reactor is generating energy at the rate of 320 megawatts. Calculate the number of U^{235} atoms undergoing the fission process. Average energy released in each fission is 200 MeV.

Solution: The amount of energy released per second in the nuclear reactor = 320 $\times 10^6$ J.

The amount of energy released per fission = 200 $\times 10^6$ $\times$ 1.602 $\times 10^{-19}$ = 3.204 $\times 10^{-11}$ J

Thus, the number of atoms undergoing fission per second is

$$= \frac{320 \times 10^6}{3.204 \times 10^{-11}} \approx 10^{19}$$

Exercise 2: A city requires on average 200 megawatts of electrical power, which is being generated by a nuclear reactor using U^{235}. The efficiency of the reactor is 30 %. Calculate the amount of U^{235} required per day. Given that the energy generated per nuclear fission is 200 MeV. Avogadro number is 6.025 $\times 10^{23}$.

Solution: Power requirement per day is $200 \times 10^6 \times 24 \times 60 \times 60 = 1.728 \times 10^{13}$ J.

Suppose x gm is the amount of U^{235} required per day. The number of U^{235} atoms in x gm is

$$= \frac{6.025 \times 10^{23} \times x}{235}$$

Due to 30% efficiency, the number of U^{235} atoms participating in the fission process is

$$= \frac{6.025 \times 10^{23} \times x}{235} \frac{30}{100} = 7.69 \times 10^{20} x$$

Thus, the energy produced is

$$= 7.69 \times 10^{20} x \times 200 \times 10^6 \times 1.602 \times 10^{-19} = 2.46 \times 10^{10} x \text{ J}$$

This amount of energy should be equal to the amount required. Thus, we have

$$2.46 \times 10^{10} x = 1.728 \times 10^{13} \qquad \text{or} \qquad x = 702.4 \text{ gm}$$

Thus, 702.4 gm of U^{235} is required per day for the electrical power supply.

Exercise 3: A nuclear reactor is generating energy at the rate of 3 megawatts. Calculate the number of U^{235} atoms undergoing the fission process per second. Calculate the amount of U^{235} which would be sufficient for running the reactor for 1000 hours. The average energy released per fission of U^{235} is 200 MeV.

Solution: Energy generated is 3×10^6 J/s

Energy released per fission $= 200 \times 10^6 \times 1.602 \times 10^{-19} = 3.204 \times 10^{-11}$ J

Thus, the number of atoms under going fission per second is

$$= \frac{3 \times 10^6}{3.204 \times 10^{-11}} = 9.36 \times 10^{16}$$

Number of atoms undergoing fission in 1000 hours is

$$= (9.36 \times 10^{16}) \times (1000 \times 60 \times 60) = 3.3696 \times 10^{23}$$

According to Avogadro hypothesis, mass of 6.025×10^{23} atoms of U^{235} is 235 gm. Thus, the mass of 3.3696×10^{23} atoms of U^{235} is

$$= \frac{235 \times 3.3696 \times 10^{23}}{6.025 \times 10^{23}} = 131.4 \text{ gm}$$

Thus, for running the reactor for 1000 hours, the required amount of U^{235} is 131.4 gm.

Exercise 4: In the fission of ${}_{92}U^{235}$ by a thermal neutron, the fragments found are ${}_{42}Mo^{98}$ and ${}_{54}Xe^{136}$. How many electrons are released in the reaction. Calculate the amount of energy released in the reaction when the masses of ${}_{92}U^{235}$, ${}_{42}Mo^{98}$, ${}_{54}Xe^{136}$ and neutron in amu are, respectively, 235.044, 97.906, 135.907 and 1.009.

Solution: For the fission of ${}_{92}U^{235}$ by thermal neutron, the reaction may be expressed as

$${}_{92}U^{235} + {}_{0}n^1 \rightarrow ({}_{92}U^{236})^* \rightarrow {}_{42}Mo^{98} + {}_{54}Xe^{136} + 2\ {}_{0}n^1 + 4 - 1e^0 + Q$$

The conservation of charge shows that four electrons should be produced in the fission process. The mass of electron is taken to be negligible in comparison to other masses.

Initial mass $= 235.044 + 1.009 = 236.053$ amu

Final mass = 97.906 + 135.907 + 2 × 1.009 = 235.831 amu

Thus, the mass loss during the fusion process = 236.053 − 235.831 = 0.222 amu

Hence, the energy released per fusion = 0.222 × 931.3 = 206.75 MeV

1.4 Fission chain reaction

In the fission of a nucleus of ${}_{92}U^{235}$ by a neutron, the fission products has about 2.5 neutrons on an average. When these neutrons have sufficient energy, they can participate in the fission of other nuclei of ${}_{92}U^{235}$. Now, about 2.5 × 2.5 = 6.25 neutrons on average would be produced. This process is shown in Figure 4, where we have taken production of three neutrons per fission. These neutrons would be used in the fission of other nuclei. Thus, a chain reaction of fission process would be developed. The number of nuclei participating in the fission would go on increasing rapidly. This process would last till all the nuclei of ${}_{92}U^{235}$ are fissioned out.

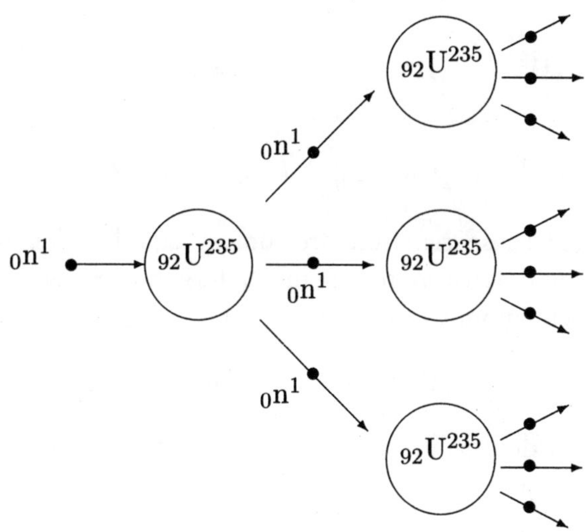

Figure 4: Chain reaction in fission.

In a similar way, as the fission of ${}_{92}U^{235}$, other nuclei showing fission process would go under the chain reaction. However, when the energy of a neutron in the fission product is not sufficient enough to produce the fission, the chain reaction would not continue.

2. Nuclear fusion

Nuclear fusion may be considered as a process which is reverse to the process of nuclear fission, as in the fusion two lighter mass nuclei combine together to from a heavier mass nucleus. In the fusion also, an amount of energy is liberated. The liberation of energy in the fission is due to the fact that the binding energy per nucleon in heavy elements is less than that of the intermediate mass numbers. The binding energy curve shows that the binding energy per nucleon in light nuclei is less than that for the nuclei of intermediate mass numbers. Hence, the sum of the masses of individual light nuclei is more than that of the nucleus formed by the fusion process. It means that fusion of two or more light nuclei will result in the liberation of energy and this will take place up to the limit $(A_1 + A_2) < 60$, where A_1 and A_2 are the masses of individual light nuclei.

Packing fraction curve indicates that the larger amount of energy would be liberated when A_1 and A_2 both are less than 12. Thus, the release of energy due to low value of binding energy per nucleon for the light nuclei is the cause of the release of energy in the fusion process. Let us consider some examples of fusion process. In case of the deuteron-deuteron reaction:

$$ {}_1\mathrm{H}^2 + {}_1\mathrm{H}^2 \rightarrow {}_1\mathrm{H}^3 + {}_1\mathrm{H}^1 + 4.03 \text{ MeV} $$

or

$$ {}_1\mathrm{H}^2 + {}_1\mathrm{H}^2 \rightarrow {}_2\mathrm{He}^3 + {}_0\mathrm{n}^1 + 3.27 \text{ MeV} $$

the cross sections for these reaction are almost equal. The reaction between deuterium and tritium has a much larger cross section and releases a much larger amount of energy:

$$ {}_1\mathrm{H}^2 + {}_1\mathrm{H}^3 \rightarrow {}_2\mathrm{He}^4 + {}_0\mathrm{n}^1 + 17.6 \text{ MeV} $$

The reaction between deuterium and ${}_2\mathrm{He}^3$ also yields a larger amount of energy but with a small cross section:

$$ {}_1\mathrm{H}^2 + {}_2\mathrm{He}^3 \rightarrow {}_2\mathrm{He}^4 + {}_1\mathrm{H}^1 + 18.4 \text{ MeV} $$

One point to be noted is that these reactions are between a pair of positively charged nuclei, which repel to each other due to Coulomb repulsion. One of the methods to bring about these reactions would be to start with a fairly large amount of deuterium, or deuterium mixed with tritium, and to increase the temperature of the mixture to a very high value. An idea about the magnitude of the required temperature can be obtained from the fact that the above reactions take place when the deuteron are accelerated to energies greater than 10 KeV. As we know that 1 eV energy is equivalent to about 12000 K temperature. Thus, 1 KeV = 12 $\times 10^6$ K. It shows that temperatures of the order of million Kelvin are required to bring about the process of nuclear fusion. Thus, the nuclear fusion reactions are also known as the thermonuclear reactions. In the nature, the fission process is considered undergoing inside the stars.

2.1 Source of stellar energy

Calculations show that our sun radiates energy at the rate of about 10^{26} J/s. The sun can be taken as a typical star whose interior is at a temperature of about 20 $\times 10^6$ K. The age of the sun is at least 5×10^9 years, so that total loss of energy during this period is exceedingly high. A question about the source of energy in the sun naturally arises. The gravitational energy of the sun has been found insufficient to produce such large amount of energy.

The fission of heavy elements cannot be the source of solar energy, because the abundance of heavy elements in the sun is too small to account for the rate of emission of energy over the life-time of the sun. It is well know that light elements have quite large abundances inside the sun, and hydrogen and helium together form about 90% by weight of the sun's matter. It seems likely, from the chemical composition, that the nuclear processes should involve hydrogen and helium and the nuclear properties of these elements support this idea.

In 1939, H A Bethe proposed that the production of energy in the sun (as well as in other shining stars) is due to thermonuclear reactions. He proposed two sets of thermonuclear reactions: (i) This reaction is known as the carbon nitrogen cycle, where the carbon and nitrogen act as the catalyst in the fusion reaction. (ii) This reaction is known as the proton-proton chain, where four protons fuse together to form a helium nucleus. Let us look into these two sets of reactions undergoing in the core of stars.

Carbon-nitrogen cycle

This set of reactions can be started either from ${}_6C^{12}$ or from ${}_7N^{14}$, as both of them work just as the catalyst. Here, we are starting from ${}_6C^{12}$. The set of reactions can be expressed as the following.

$$ {}_6C^{12} + {}_1H^1 \rightarrow {}_7N^{13} $$

$$_7N^{13} \rightarrow \; _6C^{13} + \beta^+ + \nu$$

$$_6C^{13} + \; _1H^1 \; \rightarrow \; _7N^{14}$$

$$_7N^{14} + \; _1H^1 \; \rightarrow \; _8O^{15}$$

$$_8O^{15} \rightarrow \; _7N^{15} + \beta^+ + \nu$$

$$_7N^{15} + \; _1H^1 \; \rightarrow \; _6C^{12} + \; _2He^4$$

Notice that these six reactions can be grouped into two sets, one of the first three reactions and the second of last three reactions. When we write the second set first, we get a starting from $_7N^{14}$. On addition, of these six reactions, we get a net reaction as the following.

$$4 \; _1H^1 \rightarrow \; _2He^4 + 2\beta^+ + 2\nu + Q \text{ MeV}$$

where Q is the amount of energy released per cycle. For this cycle, an important point is to know from where the $_6C^{12}$ is obtained. It is obtained through nuclear synthesis of three $_2He^4$ nuclei (as discussed later on). Though this reaction initiates at a temperature of about 10^8 K, but in some part of the core of the sun, this value of temperature may take place.

Proton-proton chain

In this chain, two protons combine to form a deuteron nucleus as the following.

$$_1H^1 + \; _1H^1 \; \rightarrow \; _1H^2 + \beta^+ + \nu + \gamma$$

The deuteron nucleus combines with another proton as the following.

$$_1H^2 + \; _1H^1 \; \rightarrow \; _2He^3 + \gamma$$

Two nuclei of $_2He^3$ so formed fuse together for form $_2He^4$ as the following.

$$_2He^3 + \; _2He^3 \; \rightarrow \; _2He^4 +_1 H^1 + +_1H^1 + \gamma$$

So, by addition of above reactions, the net effect is as the following.

$$4 \; _1H^1 \rightarrow \; _2He^4 + 2\beta^+ + 2\gamma + 27 \text{ MeV}$$

It may be noted that in this chain also, the net result is the same as that of the carbon-nitrogen cycle. Another set of reactions for the proton-proton chain is as the following.

$$_1H^1 + \; _1H^1 \; \rightarrow \; _1H^2 + \beta^+ + \gamma$$

$$_1H^1 + {}_1H^2 \rightarrow {}_2He^3 + \gamma$$

$$_2He^3 + {}_2He^3 \rightarrow {}_2He^4 + _1 H^1 + + _1H^1 + \gamma$$

$$_2He^3 + {}_2He^4 \rightarrow {}_4Be^7 + \gamma$$

$$_4Be^7 + {}_1H^1 \rightarrow {}_5B^8 + \gamma$$

$$_5B^8 \rightarrow {}_4Be^8 + \beta^+ + \gamma$$

$$_4Be^8 \rightarrow {}_2He^4 + {}_2He^4$$

Since fusion of lighter nuclei undergoes at relatively low temperature, it is argued that the former set of proton-proton chain reactions is predominates at low temperatures, whereas at the higher temperatures, the contribution of the later series becomes significant. The relative probability of carbon-nitrogen cycle and proton-proton chain reaction depends on the temperature in the core of the sun. The proton-proton chain reaction predominates up to 15 million Kelvin, while the carbon-nitrogen cycle predominates above 20 million Kelvin. Most of the energy in our sun is produced through the proton-proton chain.

Exercise 5: Suppose a helium nucleus is formed through the fusion of two deuterium nuclei. Calculate the amount of energy released. Given, mass of $_1H^2$ = 2.014102 amu, mass of $_2He^4$ = 4.002604 amu.

Solution: For the nuclear reaction

$$2\ {}_1H^2 \rightarrow {}_2He^4 + Q$$

Initial mass = $2 \times 2.014102 = 4.028204$ amu

Final mass = 4.002604 amu

Thus, the mass loss during the fusion process = $4.028204 - 4.002604 =$ 0.0256 amu

Hence, the energy released per fusion = $0.0256 \times 931.3 = 23.84$ MeV

Exercise 6: For the nuclear reaction

$$4\ {}_1H^1 \rightarrow {}_2He^4 + 2\beta^+ + 2\nu + Q$$

the masses of $_1H^1$, $_2He^4$ and β^+ are, respectively, 1.007825, 4.002603 and 0.000549 amu. Considering the neutrinos ν to be mass-less, calculate the amount of energy released per reaction.

Solution: Initial mass $= 4 \times 1.007825 = 4.0313$ amu

Final mass $= 4.002603 + 2 \times 0.000549 = 4.003701$ amu

Thus, the mass loss during the fusion process $= 4.0313 - 4.003701 = 0.027599$ amu

Hence, the energy released per fusion $= 0.027599 \times 931.3 = 25.70$ MeV

Exercise 7: The fusion reaction $2\ {}_1\text{H}^2 \rightarrow {}_2\text{He}^4 + Q$ is proposed for the production of industrial power. Assuming the efficiency of the process to be 30%, calculate how many grams of deuterium will be consumed in a day for an output of 50 megawatt. Given, mass of ${}_1\text{H}^2 = 2.014102$ amu and mass of ${}_2\text{He}^4 = 4.002604$ amu.

Solution: The mass defect per reaction $= 2 \times 2.014102 - 4.002604 = 0.0256$ amu

Thus, the energy produced for each reaction $= 0.0256 \times 931.3 = 23.84128$ MeV

Power requirement per day is $50 \times 10^6 \times 24 \times 60 \times 60 = 4.32 \times 10^{12}$ J.

Suppose x gm is the amount of ${}_1\text{H}^2$ required per day. The number of ${}_1\text{H}^2$ atoms in x gm is

$$= \frac{6.025 \times 10^{23} \times x}{2}$$

Due to 30% efficiency, the number of ${}_1\text{H}^2$ atoms participating in the fission process is

$$= \frac{6.025 \times 10^{23} \times x}{2} \frac{30}{100} = 9.0375 \times 10^{22} x$$

Thus, the energy produced is

$$= 9.038 \times 10^{22} x \times 23.84128 \times 10^6 \times 1.602 \times 10^{-19} = 3.45 \times 10^{11} x \text{ J}$$

This amount of energy should be equal to the amount required. Thus, we have

$$3.45 \times 10^{11} x = 4.32 \times 10^{12} \qquad \text{or} \qquad x = 12.522 \text{ gm}$$

Thus, 12.522 gm of ${}_1\text{H}^2$ is required per day for the power supply.

Exercise 8: Calculate the amount of energy produced for complete fission of 23.5 gm of ${}_{92}\text{U}^{235}$. For the fission of one nucleus of ${}_{92}\text{U}^{235}$, the energy produced is 200 MeV.

Solution: According to Avogadro hypothesis, mass of 6.025×10^{23} atoms of U^{235} is 235 gm. Thus, the number of atoms in 23.5 gm of $_{92}U^{235}$ is

$$= \frac{6.025 \times 10^{23} \times 23.5}{235} = 6.025 \times 10^{22}$$

Thus, total energy produced from 23.5 gm of $_{92}U^{235} = 6.025 \times 10^{22} \times 200 = 1.205 \times 10^{25}$ MeV

2.2 Journey of nuclear synthesis up to $A = 56$.

We have learned that in the first phase of shining of a young star, burning (fusion) of protons into He^4 takes place. When most of the protons in a young star are converted into He^4, the production of fusion energy drops significantly and the star begins to cool down due to the loss of energy through radiation to the surrounding space. With the drop of the temperature, thermal pressure inside the star reduces and the process of gravitational contraction under its own gravity resumes again and continues until the star is heated up to a temperature higher than that was required for the fusion (burning) of protons. At the temperature of about 10^8 K, corresponding to the kinetic energy of the order of 10 KeV, collisions between He^4 nuclei become sufficiently energetic, so that the nuclear reactions involving He^4 become significant. This starts the burning of He^4.

The basic reaction in the burning of He^4 is the conversion of three He^4 nuclei into a C^{12} nucleus, similar to the conversion of four protons into a He^4 nucleus during the burning of protons. However, the probability of a three-body reaction for fusion of three He^4 nuclei directly into a C^{12} is too low. Hence, the formation of C^{12} occurs in two stages. The first one is the formation of a Be^8 nucleus from two He^4 nuclei through the process

$$He^4 + He^4 \rightarrow Be^8$$

Although Be^8 is unstable, lifetime of 2.6×10^{-16} s, but in the ground state, it is sufficiently large time so that the reaction

$$He^4 + Be^8 \rightarrow C^{12} \tag{5.3}$$

can take place. This is partly due to the large cross-section for this reaction. When the amount of C^{12} accumulated in a star is large enough, probability of capture of He^4 by C^{12} becomes quite significant. The following reaction also has a large cross section.

$$He^4 + C^{12} \rightarrow O^{16} \tag{5.4}$$

The production rates for C^{12} in reaction (5.3) and for O^{16} in reaction (5.4) are so high that the abundance of C^{12} and O^{16} in the solar system is at the second position while the abundance of proton (H) and α-particle (He) is at the first position. Other light nuclei with α-cluster structure, such as Ne^{20} and Mg^{24}, are also formed in significant quantities by successive capture of α-particles. (Most of the observed element heavier than He^4 in the solar system are not produced inside the sun. They are primarily the remnant of the stars which have gone through their life cycles (supernova) before the formation of our solar system.)

When all the He^4 is used up, the star goes through another stage of the gravitational contraction under its own gravity and thereby increasing the temperature inside the star. (Such situation will never arrive inside the sun, but is the case inside the heavy stars.) As the temperature rises above 10^9 K, corresponding to 100 KeV in average kinetic energy per particle, then the reactions involving the conversion of C^{12} into heavier nuclei becomes possible. For example

$$\begin{aligned} C^{12} + C^{12} &\to Mg^{24} + \gamma \\ &\to Na^{23} + p \\ &\to Mg^{23} + n \\ &\to Ne^{23} + \alpha \\ &\to O^{16} + \alpha + \alpha \end{aligned}$$

At even higher temperatures, conversion of O^{16} to heavier element becomes significant through the following reactions.

$$\begin{aligned} O^{16} + O^{16} &\to S^{32} + \gamma \\ &\to P^{31} + p \\ &\to S^{31} + n \\ &\to Si^{28} + \alpha \\ &\to Mg^{24} + \alpha + \alpha \end{aligned}$$

On the other hand, the reaction $C^{12} + O^{16}$ is not considered to be important in the nuclear synthesis, because nearly all the C^{12} will be exhausted before the temperature is high enough for the reaction to become significant.

The process of formation of heavier elements successively through exothermic nuclear reactions stops around $A = 56$ (Fe^{56} and Ni^{56}). The nuclear binding energy reaches to its maximum value at this stage.

3. Problems and questions

1. Using the liquid drop model for a nucleus, explain the phenomenon of fission of a heavy nucleus.

2. For a fission process, derive an expression for critical energy on the basis of the liquid drop model. Show that a nucleus with Z^2/A around 50 would undergo fission instantaneously.

3. Describe about the source of energy in the core of our sun.

4. Distinguish between the nuclear fission and nuclear fusion.

5. Write notes on the following

   (i) Nuclear fission

   (ii) Bohr and Wheeler theory for nuclear fission

   (iii) Nuclear fusion

   (iv) Formation of heavy mass nuclei

VI. Radioactive Decay

Radioactivity was first discovered by H. Becquerel in 1896. He noticed that a uranium salt wrapped in a paper emitted some kind of radiation which affected the photographic plate, placed nearby. Further investigation revealed that emission of these radiation did not require any kind of energy or force from out side which would have been used for some type of excitation. It was also found that there was no effect of any external source (magnetic field etc.) on these radiation.[1] Detailed study of such radiation was further carried out by Madame Curie, Pierre Curie, Rutherford and others. It was also found that such radiation are emitted only by heavy elements, like uranium, polonium, radium, thorium.

During the study of various materials, in 1899, Rutherford noticed that radiation from some materials had very different type of characteristics, as some unknown kind of radiation were present there. These radiation were found deflected in two opposite directions, in the presence of magnetic field. He called those radiation as the α-rays and β-rays. The penetration power of these radiation was different from each other. In 1900, Villard found that besides these two kinds of α and β-rays, there was third type of radiation, which could not be deflected in a magnetic field. These radiation were named as the γ-rays. The penetration power of γ-rays was found to be much larger than that of α-rays as well as of β-rays. The materials which were found emitting such type of radiation were named as the radioactive elements. The materials with atomic mass larger than 206 show the phenomenon of natural radioactivity.

With the help of artificial transmutation of elements, it is nowadays possible to produce radioactivity in other elements, which are much lighter than those occurring in nature. Such activity is sometimes is known as the induced radioactivity or artificial radioactivity, in order to distinguish it from the natural radioactivity. Whatever be the type of radioactivity, it is a spontaneous phenomenon.

Further studies showed that α-rays and β-rays are not so strictly

[1]The noticed effect would have been due to γ-radiation. By that time, nothing was known about the α and β-particles.

radiation. They were particles having finite mass. So, they are also known as α-particles and β-particles, respectively. On the basis of the directions of the deflections in a magnetic field, it was found that α-particles carry positive charge, β-particles carry negative charge and γ-rays have no charge. These radiation (α-particles, β-particles and γ-rays) are also known as the Becquerel rays. Becquerel proved that the production of these rays is an atomic phenomenon and not a molecular one. Actually, it is the nuclear phenomenon.

The radioactive decay is always accompanied by a net loss of mass. This mass loss appears in the form of equivalent amount of energy. The nucleus which undergoes the radioactive decay (*i.e.*, the emission of α-particles, β-particles and γ-rays) is known as the parent (or mother) nucleus. The nucleus, which is formed after the radioactive decay, is known as the daughter nucleus.

1. Alpha, beta and gamma rays

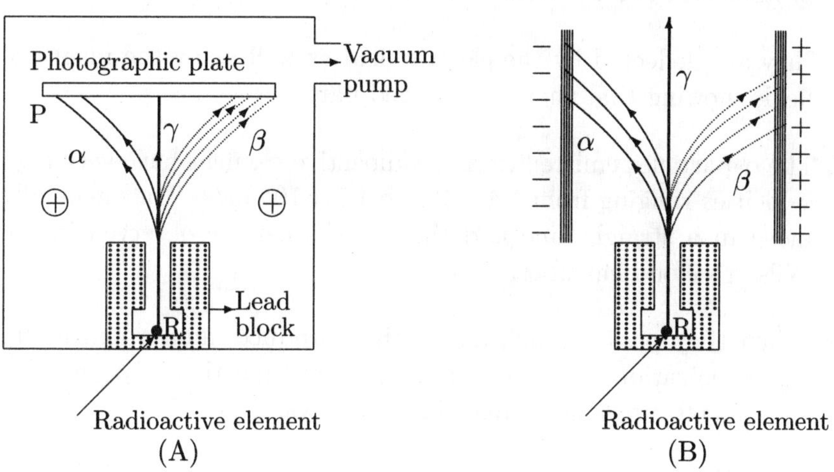

Figure 1: Experiments showing that the emission of α-particle, β-particle and γ-radiation by a radioactive element.

The fact that the radiation emitted by radioactive elements are of three types, called α-particles, β-particles and γ-rays, can be shown with the help of simple experiments (Figure 1). A small quantity of radium R is placed at the bottom of a small hole drilled in a lead block. A fairly parallel beam of radiation from R will come out though the hole. The lead block is generally placed inside an evacuated chamber, in order to avoid the absorption of the rays.

In one experiment, shown in Figure 1 (A), we apply a strong magnetic field perpendicular to the plane of paper in the inward direction. A photographic plate P is placed at a short distance above the lead block to receive the rays. After a fairly long exposure, the photographic plate would show three distinct marks, corresponding to three different types of rays. The α-particles were deflected towards the left, indicating that they carry positive charge. The β-particles were deflected towards the right, indicating that they carry negative charge. The γ-rays were not deflected, indicating that they carry no charge.

In another experiment, shown in Figure 1 (B), we apply an electrostatic field. The α-particles are deflected towards the negatively charged plate whereas the β-particles are deflected towards the positively charged plate. The γ-rays again do not show any deflection.

1.1 Properties of α-particles

(i) An α-particle is, in principle, a helium nucleus consisting of two protons and two neutrons. Thus, it has mass 4 and charge +2.

(ii) They are deflected by the electric field as well as by the magnetic field, showing that they are charged particles.

(iii) The α-particles emitted from a radioactive element have very high velocities ranging from 1.4×10^7 to 1.7×10^7 m/s. They generally move in a straight line and their tracks can be observed in the Wilson's cloud chamber.

(iv) When they pass through a gas, they produce intense ionization. Their ionization power is 100 times more than that of β-particles and 10,000 times more than that of γ-rays.

(v) They affect a photographic plate, but the effect is very feeble.

(vi) They penetrate through thin metal foils and their penetration power is smaller than that of β-particles.

1.2 Properties of β-particles

(i) A β-particle is, in principle, an electron having charge 1 and rest mass of 9.1×10^{31} kg.[2]

---

[2]Quite often, a positron (anti-particle of electron) is also called a β-particle. Then, we classify as β^- and β^+, for electron and positron, respectively.

(ii) They are deflected by the electric field as well as by the magnetic field, showing that they are charged particles. The direction of deflection in the case of a β-particle is opposite to that α-particle, showing that α and β-particles carry opposite charges.

(iii) Their ionization power is low and therefore the range is large.

(iv) They affect a photographic plate, but the effect is very feeble.

(v) They penetrate through thin metal foils and their penetration power is larger than that of α-particles.

1.3 Properties of γ-radiation

(i) The γ-rays are, in principle, electromagnetic waves of very short wavelength ranging from 0.005 Å to 0.05 Å.

(ii) They are neither deflected by the electric field nor by the magnetic field, showing that they do not have any charge.

(iii) They affect a photographic plate very strongly.

(iv) They ionize the gas through which they pass, but the ionization produced is very small.

(v) They are more penetrating than even β-particles.

2. Fundamental laws of radioactivity

When a radioactive disintegration takes place with the emission of α-particle and/or β-particle, the original atom (called the parent atom) changes into another atom (called the daughter atom). With the emission of γ-radiation, the nucleus remains the same. There are three fundamental laws of radioactivity as the following:

2.1 Displacement laws

In 1913, some simple laws known as the displacement laws of radioactivity were worked out and they are as the following.

(a) When a radioactive atom emits an α-particle (mass 4 and charge 2), it is converted into another element of atomic number two less and mass number 4 less than that of the parent element. Consequently, the place of the new atom is shifted two groups lower in the periodic table. The following reaction shows the emission of α-particle.

$$ {}_{92}\mathrm{U}^{238} \xrightarrow{-\alpha} {}_{90}\mathrm{Th}^{234} $$

(b) When a radioactive atom emits a β-particle (mass approximately 0 and charge -1), it is converted into another element of atomic number one greater than that of the parent element but of the same mass number. Consequently, the place of the new atom is shifted one group higher in the periodic table. The following reaction shows the emission of β-particle.

$$ {}_{90}\mathrm{Th}^{234} \xrightarrow{-\beta^-} {}_{91}\mathrm{Pa}^{234} $$

With the help of the displacement law, it is possible to determine the mass and atomic numbers of different elements in the successive radioactive disintegration, when the mass and atomic numbers of parent nucleus are known.

2.2 Law of radioactive disintegration

Based on the experimental observations, Rutherford and Soddy in 1902 found that the rate at which a particular radioactive material disintegrates or decays was independent of physical and chemical conditions and was dependent of number of atoms present at that time. Since disintegration goes on continuously, the number of atoms present keeps on changing, hence the rate of disintegration will change with time. All the radioactive decays are spontaneous and obey the simple kinetic laws but follow different mechanisms.

As per observations, the rate of decay or disintegration of a radioactive element at any instant t is proportional to the number of atoms N present at that instant, that is

$$ -\frac{\mathrm{d}N}{\mathrm{d}t} \propto N \qquad \text{or} \qquad -\frac{\mathrm{d}N}{\mathrm{d}t} = \lambda N \tag{6.1} $$

where λ is a constant, known as the disintegration constant or the decay constant of the radioactive element. It depends on the element. Negative sign indicate that number of atoms of the radioactive element decreases with time. Solution of equation (6.1) is

$$ N = N_0 \, \mathrm{e}^{-\lambda t} \tag{6.2} $$

where N_0 is the initial number of the atoms of the radioactive element. Here, we have assumed that the probability of disintegration per second

λ is independent of the age of that atom and is the same for all the atoms of the element. For a nucleus having several modes of decay, λ is the total probability of decay and is the sum of the probabilities λ', λ'', λ''', ... of the individual modes of decay. The decay constant λ is one of the most important characteristics of each radioactive element and is independent of physical and chemical conditions such as temperature, pressure, concentration or age of the radioactive atoms. Measurement of λ can identify the radioactive samples and because of that, it is also called radioactive constant.

Equation (6.2) shows that the number of atoms of a given radioactive element decreases exponentially with time provided that no new atoms are introduced. It is shown in Figure 2.

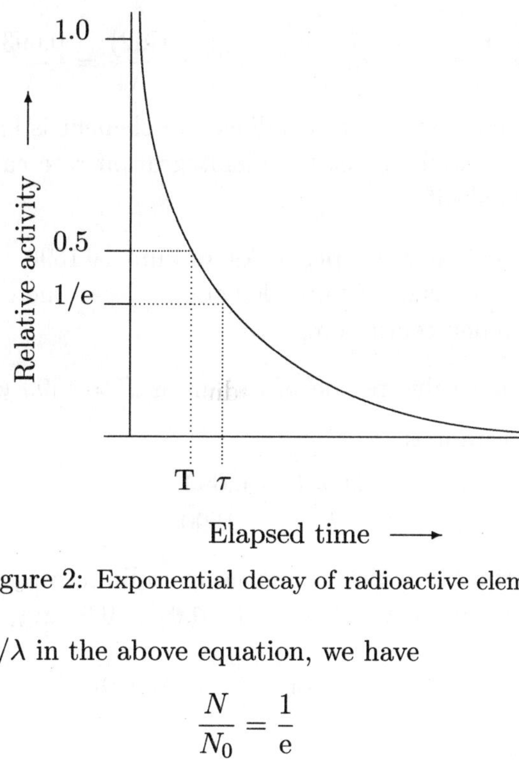

Figure 2: Exponential decay of radioactive element.

Putting $t = 1/\lambda$ in the above equation, we have

$$\frac{N}{N_0} = \frac{1}{\mathrm{e}}$$

Thus, the radioactive constant λ is the reciprocal of the time during which the number of atoms of a radioactive element falls to 1/e of its initial value. Equation (6.2) can be written as

$$\ln N = \ln N_0 - \lambda t \qquad \text{or} \qquad \log N = \log N_0 - 0.4343\ \lambda t$$

This expression shows that when logarithm of N is plotted against the time t, we shall get a straight line whose slope will be $-0.4343\ \lambda$.

Half-period

The usual way to specify how fast disintegrations of a radioactive element takes place, is in terms of the half-period, denoted by T or $t_{1/2}$. It is defined as the time required for disintegration of one half of the initial number of nuclei in the element. It is specific characteristic of the radioactive element. The half-period is sometimes known as the half-value time. If N_0 be the initial number of nuclei of radioactive element at time $t = 0$ and therefore, after the half-period T, the number of nuclei reduces to $N = N_0/2$. From the relation

$$N = N_0\ e^{-\lambda t} \qquad \text{we have} \qquad \frac{N_0}{2} = N_0\ e^{-\lambda T}$$

Thus, we have

$$e^{-\lambda T} = \frac{1}{2} \qquad \text{or} \qquad T = \frac{(\ln 2)}{\lambda} = \frac{0.693}{\lambda} \tag{6.3}$$

When the decay constant λ for a radioactive element is known, its half-life T can be calculated. Thus, the disintegration rate can be obtained with the help of half-life.

Exercise 1: The half-value period of radium is 1590 years. In how many years will one gram of pure element (i) loses one centigram, and (ii) be reduced to one centigram.

Solution: The half-value period of radium is $T = 1590$ years.

The radiative constant is

$$\lambda = \frac{0.6931}{T} = \frac{0.6931}{1590}$$

(i) Let t be the time in which one gram of radium loses one centigram (0.01 gm). The radium left behind = 1 - 0.01 = 0.99 gm. Now, we have

$$N = N_0\ e^{-\lambda t} \qquad \text{or} \qquad \lambda t = \ln\left(\frac{N_0}{N}\right)$$

Therefore, we have

$$t = \frac{1}{\lambda}\ln\left(\frac{N_0}{N}\right) = \frac{1590}{0.6931}\ln\left(\frac{1}{0.99}\right) = 23.25 \text{ years}$$

(ii) Let t be the time in which one gram of radium converts into one centigram (0.01 gm). Now, we have

$$t = \frac{1}{\lambda}\ln\left(\frac{N_0}{N}\right) = \frac{1590}{0.6931}\ln\left(\frac{1}{0.01}\right) = 10560 \text{ years}$$

Exercise 2: One gram of pure radium is reduced by 2.1 mg in 5 years through α-decay. Calculate the half-life period of radium.

Solution: Initial mass of radium $N_0 = 1$ gm

Mass of radium after 5 years $N = 1 - 0.0021 = 0.9979$ gm

With the radiative decay constant λ, we have

$$N = N_0 \, e^{-\lambda t} \qquad \text{or} \qquad \lambda = \frac{1}{t} \ln\left(\frac{N_0}{N}\right)$$

Therefore, we have

$$\lambda = \frac{1}{5} \ln\left(\frac{1}{0.9979}\right) = 41.45 \times 10^{-5} \text{ per year}$$

Thus, the half-life time of radium is

$$T = \frac{0.6931}{\lambda} = \frac{0.6931}{41.45 \times 10^{-5}} = 1672 \text{ years}$$

Exercise 3: Calculate the time-period required for 10% of thorium to disintegrate. Given, the half-life of thorium = 1.4 $\times 10^{10}$ years.

Solution: From the given half-life time $T = 1.4 \times 10^{10}$ years, the disintegration constant λ is

$$\lambda = \frac{0.6931}{T} = \frac{0.6931}{1.4 \times 10^{10}}$$

There is 10% disintegration. The reamining amount is 90%. Thus, $N = 0.9 N_0$. Now, we have

$$N = N_0 \, e^{-\lambda t} \qquad \text{or} \qquad t = \frac{1}{\lambda} \ln\left(\frac{N_0}{N}\right)$$

Therefore, we have

$$t = \frac{1.4 \times 10^{10}}{0.6931} \ln\left(\frac{1}{0.9}\right) = 2.1 \times 10^{9} \text{ years}$$

Average life-time

In a radioactive element, owing to the continuous disintegration, the number of parent nuclei changes continuously. Some atoms may have short life-time whereas the others may have long life-time. Thus, it is of interest to determine the average life-time, denoted by τ, of the parent nuclei. Obviously, with the decrease of the number of parent nuclei, the number of daughter nuclei will increase continuously. The average life-time of parent nuclei may be taken equal to the sum of all life-times of parent nuclei divided by the total number of parent nuclei. Let us assume that ΔN is the number of parent nuclei having time t. Thus, we have

$$\tau = \frac{1}{N_0} \sum t \, \Delta N$$

When ΔN is very small and the initial number of parent nuclei is N_0 which all of them finally disintegrate, then we have

$$\tau = \frac{1}{N_0} \int_0^{N_0} t \, \mathrm{d}N \tag{6.4}$$

Using N from equation (6.2) in (6.4), we get

$$\tau = -\lambda \int_0^{\infty} t \, \mathrm{e}^{-\lambda t} \, \mathrm{d}t = \frac{1}{\lambda} \tag{6.5}$$

It shows that the average life-time is inversely proportional to the disintegration constant. Thus, the element having large value for the disintegration constant has short life-time.

Relation between average life-time and half-period

In fact, both the average life-time and half-period are the properties of a common parent nucleus. Thus, it not surprising that they are related with each other. From equations (6.5) and (6.3), a relation between the average life-time τ and the half-period T can be obtained as the following.

$$\tau = \frac{T}{0.693} = 1.44 \, T$$

Exercise 4: 1 gm of radioactive element disintegrate at the rate of 3.7 $\times 10^{10}$ disintegrations per second. Calculate the mean-life time of the element. Given, atomic mass number of the element = 226.

Solution: For the element with atomic mass number 226, the number of nuclei in 1 gm of element is

$$N = \frac{6.023 \times \times 10^{23} \times 1}{226} = 2.665 \times 10^{21}$$

The disintegration rate is

$$-\frac{\mathrm{d}N}{\mathrm{d}t} = 3.7 \times 10^{10}$$

The radioactive disintegration constant is obtained as

$$\lambda = \frac{-\mathrm{d}N/\mathrm{d}t}{N} = \frac{3.7 \times 10^{10}}{2.665 \times 10^{21}} = 1.388 \times 10^{-11}\ \mathrm{s}^{-1}$$

Thus, the mean-life is

$$\tau = \frac{1}{\lambda} = \frac{1}{1.388 \times 10^{-11}} = 7.20 \times 10^{10}\ \mathrm{s} = 2283 \text{ years}$$

Exercise 5: The disintegration constant of a radioactive element is 0.00231 per day. Calculate half-life and average-life of the element.

Solution: We have $\lambda = 0.00231$ per day. Therefore, the half-life period is

$$T = \frac{0.693}{\lambda} = \frac{0.693}{0.00231} = 300 \text{ days}$$

Average life-time is

$$\tau = \frac{1}{\lambda} = \frac{1}{0.00231} = 432.9 \text{ days}$$

Units of radioactivity

Generally, the activity of radioactive elements is expressed in the units of Curie, denoted by Ci. One Curie is defined as the quantity of a radioactive element which gives 3.70×10^{10} disintegrations/second. It is the number of disintegrations per second per gram of radium.

Another unit used is the Rutherford, denoted by Rd. One Rutherford is defined as the quantity of a radioactive element which gives 10^6 disintegrations/second.

Exercise 6: Calculate the mass in gm of one Curie of radioactive element Pb^{214} whose half-life is 26.8 minutes.

Solution: For the half-life $T = 26.8 \times 60 = 1608$ s, the radioactive constant λ is

$$\lambda = \frac{0.693}{1608} = 4.31 \times 10^{-4}\ \mathrm{s}^{-1}$$

Now, 1 Curie = 3.70×10^{10} disintegrations/second

Thus, $-\mathrm{d}N/\mathrm{d}t = 3.70 \times 10^{10}$

From the relation

$$-\frac{\mathrm{d}N}{\mathrm{d}t} = \lambda N \qquad \text{we have} \qquad N = \frac{-\mathrm{d}N/\mathrm{d}t}{\lambda}$$

Thus, we get

$$N = \frac{3.70 \times 10^{10}}{4.31 \times 10^{-4}} = 8.585 \times 10^{13} \text{ atoms}$$

According to Avogadro hypothesis, 6.023 $\times 10^{23}$ atoms are contained in 214 gm. Thus, the mass of 8.585 $\times$ 10^{13} atoms of Pb^{214} is

$$= \frac{214 \times 8.585 \times 10^{13}}{6.023 \times 10^{23}} = 3.1 \times 10^{-8} \text{ gm}$$

Radioactive dating

The decay of radioactive elements is independent of physical and chemical conditions imposed on them. A radioactive substance goes on decaying with time and therefore the knowledge of the number of nuclei can help in knowing about the passage of time. Here, we shall discuss two cases: (i) Age of the earth and (ii) Age of the plant.

(i) Age of the earth

The age of the earth has been estimated from the relative abundances of two isotopes of uranium, U^{238} and U^{235}. The half-periods of U^{238} and U^{235} are respectively 4.5 $\times 10^9$ years and 7 $\times 10^8$ years. For the estimation of the age of the earth, we assume that at the time of its formation, the two isotopes of uranium would have been in equal proportion. The number of atoms of the two isotopes today can be expressed as

$$N_1 = N_1(0)\, e^{-\lambda_1 t} \qquad \text{and} \qquad N_2 = N_2(0)\, e^{-\lambda_2 t}$$

At the time of formation of earth, we have assume $N_1(0) = N_2(0)$ and therefore the ratio is

$$\frac{N_1}{N_2} = e^{(\lambda_2 - \lambda_1)t} \qquad \text{or} \qquad t = \frac{1}{(\lambda_2 - \lambda_1)} \ln\left(\frac{N_1}{N_2}\right)$$

Now,

$$\lambda_1 = \frac{0.693}{4.5 \times 10^9} = 1.54 \times 10^{-10} \text{ year}^{-1}$$

and

$$\lambda_2 = \frac{0.693}{7 \times 10^8} = 9.9 \times 10^{-10} \text{ year}^{-1}$$

and U^{238} and U^{235} are found as 99.3% and 0.7%. Thus, we have

$$t = \frac{1}{(9.9 \times 10^{-10} - 1.54 \times 10^{-10})} \ln\left(\frac{99.3}{0.7}\right) = 5.93 \times 10^9 \text{year}$$

Exercise 7: The half-periods of two isotopes A and B of a radioactive substance are, respectively, 2.31×10^9 years and 3.465×10^8 years. Assuming that at the time of formation of earth, A and B were in the ratio of 1 : 2, calculate the age of earth when their present ratio is 98 : 2.

Solution: The number of atoms of the two isotopes A and B today can be expressed as

$$N_A = N_A(0)\ \mathrm{e}^{-\lambda_A t} \qquad \text{and} \qquad N_B = N_B(0)\ \mathrm{e}^{-\lambda_B t}$$

At the time of formation of earth, we have assume $N_A(0) = N_B(0)/2$ and therefore the ratio is

$$\frac{N_A}{N_B} = \frac{1}{2}\ \mathrm{e}^{(\lambda_B - \lambda_A)t} \qquad \text{or} \qquad t = \frac{1}{(\lambda_B - \lambda_A)} \ln\left(\frac{2N_A}{N_2}\right)$$

Now, we have

$$\lambda_A = \frac{0.693}{2.31 \times 10^9} = 3 \times 10^{-10} \text{ year}^{-1}$$

and

$$\lambda_B = \frac{0.693}{3.465 \times 10^8} = 2 \times 10^{-9} \text{ year}^{-1}$$

Thus, we have

$$t = \frac{1}{(2 \times 10^{-9} - 3 \times 10^{-10})} \ln\left(\frac{2 \times 98}{2}\right) = 2.70 \times 10^9 \text{ year}$$

(ii) Age of the plant

All the plants consume CO_2 from the atmosphere. A part of carbon in the plants is C^{14} which is radioactive. Hence, the plants are slightly radioactive. When a plant dies, no additional C^{14} is taken in and that within the plant body begins to decay without being replaced. Measurement of relative amounts of C^{14} and C^{12} in an organic archaeological sample provides a sensitive method of dating.

Exercise 8: A carbon specimen found in a cave contained 1/8 as much C^{14} as an equal amount of carbon in living matter. Calculate the approximate age of the specimen. Half-life period of C^{14} is 5568 years.

Solution: The half-life period $T = 5568$ years. Therefore the radioactive constant is

$$\lambda = \frac{0.693}{T} = \frac{0.693}{5568} = 1.2446 \times 10^{-4} \text{ year}^{-1}$$

Now, we have $N/N(0) = 1/8$. From the relation

$$N = N(0)\ e^{-\lambda t} \qquad \text{we have} \qquad t = \frac{1}{\lambda}\ \ln\left[\frac{N(0)}{N}\right]$$

Using he values, we get

$$t = \frac{1}{1.2446 \times 10^{-4}}\ \ln(8) = 16,708 \text{ years}$$

2.3 Law of successive transformation

In the successive transformation, one radioactive element decays into another, which is also radioactive. The first one is called the parent (or mother) element and the second is known as the daughter element. This relation is not limited to the parent and daughter, but extends over to further generations, until a stable end-product is reached. Experiments have shown that the naturally occurring radioactive elements form four series as a function of time t.

Natural radioactive series

In the successive disintegration process, where radioactive element is produced at each stage, various elements form a series. The first member of the series is known as the parent, the intermediate members are known as the daughters and final stable nucleus is known as the end-product. Four such natural series, starting from uranium, actinium, thorium and neptunium are known at present. Three of them are as the following. In each of these series, the final product is some isotope of the stable nucleus $_{82}$Pb. Each of them is consists of a succession of daughter elements.

Uranium series:

$$U^{238} \xrightarrow{\alpha} Th^{234} \xrightarrow{\beta} Pa^{234} \xrightarrow{\beta} U^{234} \xrightarrow{\alpha} Th^{230} \xrightarrow{\alpha} Ra^{226}$$

$$Ra^{226} \xrightarrow{\alpha} Rn^{222} \xrightarrow{\alpha} Po^{218} \xrightarrow{\alpha} Pb^{214} \xrightarrow{\beta} Bi^{214} \xrightarrow{\beta} Po^{214}$$

$$Po^{214} \xrightarrow{\alpha} Pb^{210} \xrightarrow{\beta} Bi^{210} \xrightarrow{\beta} Po^{210} \xrightarrow{\alpha} Pb^{206}\ \text{(stable)}$$

Actinium series:

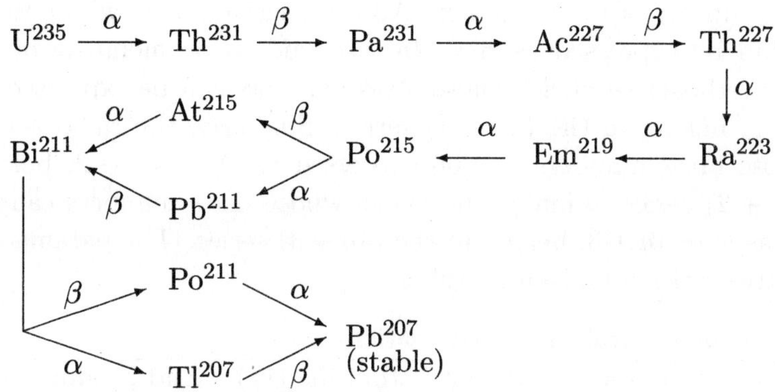

Thorium series:

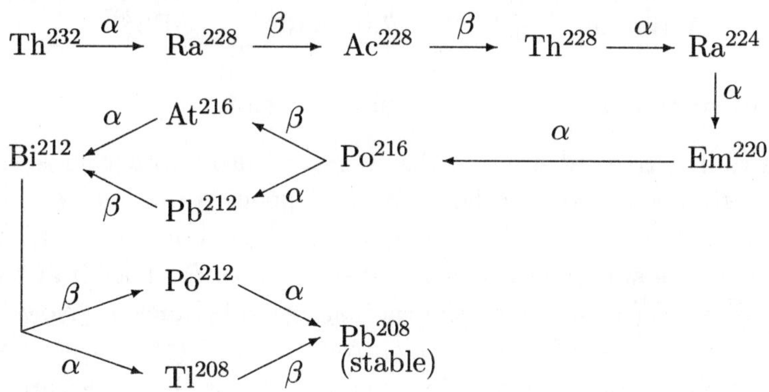

Neptunium series:

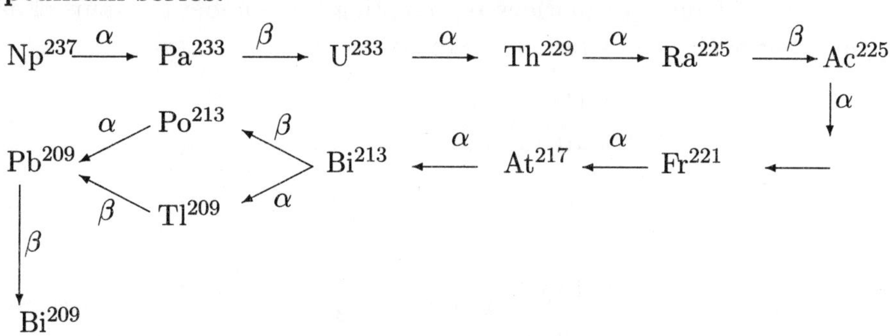

An explanation may be given as to why there are four natural series. We are aware of the facts that an α-decay reduces the mass number by 4. (Emission of β-particle as well as γ-radiation does not change the mass number of the nucleus.) Thus, a nucleus whose mass number is

$A = 4n$, where n is a positive integer, can decay into one another in descending number of mass number. All radioactive nuclei whose mass numbers can be expressed as $A = 4n$, are said to be members of $4n$ series. The radioactive nuclei whose mass numbers can be expressed as $A = 4n + 1$, belong to the $(4n + 1)$ series. Similarly, the radio active nuclei whose mass numbers can be expressed as $A = 4n + 2$, belong to the $(4n + 2)$ series. Finally, the nuclei whose mass numbers can be expressed as $A = 4n + 3$, belong to the $(4n + 3)$ series. The parameters of radioactive series are given in Table 1.

Table 1. Four natural radioactive series

Mass No.	Series	Parent	Half-life (yr)	End-product
$4n$	Thorium	$_{90}Th^{232}$	1.39×10^{10}	$_{82}Pb^{208}$
$4n + 1$	Neptunium	$_{93}Np^{237}$	2.25×10^{6}	$_{83}Bi^{209}$
$4n + 2$	Uranium	$_{92}U^{238}$	4.51×10^{9}	$_{82}Pb^{206}$
$4n + 3$	Actinium	$_{92}U^{235}$	7.07×10^{8}	$_{82}Pb^{207}$

Number of nuclei in successive disintegration

Suppose a radioactive substance 1 disintegrates into another substance 2. This substance 2 also is radioactive and disintegrates into another substance 3. This process of disintegration would continue until the product nucleus is stable and does not disintegrates further. If at time $t = 0$, we have $N_1(0)$ atoms of the radioactive substance 1. Now, we have to find the number of atoms, $N_1(t)$, $N_2(t)$, ..., $N_n(t)$ present at any subsequent time t. Suppose, the decay constants for the nuclei are λ_1, λ_2, ..., λ_n, respectively. Obviously, all the product nuclei $N_2(t)$, $N_3(t)$, ..., $N_n(t)$ all are zero at $t = 0$.

The differential equations representing the number of atoms of each member of the series can be expressed as

$$\frac{dN_1}{dt} = -\lambda_1 N_1 \tag{6.6}$$

$$\frac{dN_2}{dt} = \lambda_1 N_1 - \lambda_2 N_2 \tag{6.7}$$

$$\frac{dN_3}{dt} = \lambda_2 N_2 - \lambda_3 N_3 \tag{6.8}$$

$$\frac{dN_4}{dt} = \lambda_3 N_3 - \lambda_4 N_4 \tag{6.9}$$

$$\vdots \qquad \vdots \qquad \vdots$$

$$\frac{dN_n}{dt} = \lambda_{n-1} N_{n-1} - \lambda_n N_n$$

From equation (6.6), the number of atoms N_1 can be written as

$$N_1 = N_1(0)\ \mathrm{e}^{-\lambda_1 t} \tag{6.10}$$

This expression gives the number of atoms of substance 1 at time t. Now, the number of atoms of substance 2, increases because of the decay of the substance 1 into the substance 2, and decreases because of the decay of the substance 2 into the substance 3. Inserting this value of N_1 from equation (6.10) in (6.7) and rearranging, we have

$$\frac{\mathrm{d}N_2}{\mathrm{d}t} + \lambda_2 N_2 = \lambda_1 N_1(0)\ \mathrm{e}^{-\lambda_1 t}$$

Multiplying this equation throughout by $\mathrm{e}^{\lambda_2 t}$ and then integrating, we have

$$N_2\ \mathrm{e}^{\lambda_2 t} = \frac{\lambda_1 N_1(0)}{\lambda_2 - \lambda_1}\ \mathrm{e}^{(\lambda_2 - \lambda_1)t} + C' \tag{6.11}$$

where C' is the constant of integration. Since in the beginning, only the parent substance 1 was present. Thus, at $t = 0$, we have $N_2 = 0$ and therefore from equation (6.11), we get

$$C' = -\frac{\lambda_1 N_1(0)}{\lambda_2 - \lambda_1}$$

Substituting value of C' in equation (6.11) and multiplying throughout by $\mathrm{e}^{-\lambda_2 t}$, we get

$$N_2 = \frac{\lambda_1 N_1(0)}{(\lambda_2 - \lambda_1)}\left[\mathrm{e}^{-\lambda_1 t} - \mathrm{e}^{-\lambda_2 t}\right] \tag{6.12}$$

This expression gives the number of atoms of substance 2 at time t. Now, the number of atoms of substance 3 decreases because the substance 3 decays into the substance 4, and increases because of the decay of the substance 2 into the substance 3. Inserting this value of N_2 from equation (6.12) in (6.8), we have

$$\frac{\mathrm{d}N_3}{\mathrm{d}t} + \lambda_3 N_3 = \frac{\lambda_1 \lambda_2 N_1(0)}{(\lambda_2 - \lambda_1)}\left[\mathrm{e}^{-\lambda_1 t} - \mathrm{e}^{-\lambda_2 t}\right]$$

Multiplying this equation throughout by $\mathrm{e}^{\lambda_3 t}$ and then integrating, we have

$$N_3\ \mathrm{e}^{\lambda_3 t} = \frac{\lambda_1 \lambda_2 N_1(0)}{(\lambda_2 - \lambda_1)}\left[\frac{\mathrm{e}^{(\lambda_3 - \lambda_1)t}}{(\lambda_3 - \lambda_1)} - \frac{\mathrm{e}^{(\lambda_3 - \lambda_2)t}}{(\lambda_3 - \lambda_2)}\right] + C'' \tag{6.13}$$

where C'' is the constant of integration. In the beginning, only the parent substance 1 was present. Thus, at $t = 0$, we have $N_3 = 0$ and therefore from equation (6.13), we get

$$C'' = -\frac{\lambda_1\lambda_2 N_1(0)}{(\lambda_2 - \lambda_1)}\left[\frac{1}{(\lambda_3 - \lambda_1)} - \frac{1}{(\lambda_3 - \lambda_2)}\right]$$

Substituting the value of C'' in equation (6.13) and multiplying throughout by $e^{-\lambda_3 t}$, we get

$$N_3 = \frac{\lambda_1\lambda_2 N_1(0)}{(\lambda_2 - \lambda_1)}\left[\frac{e^{-\lambda_1 t} - e^{-\lambda_3 t}}{(\lambda_3 - \lambda_1)} - \frac{e^{-\lambda_2 t} - e^{-\lambda_3 t}}{(\lambda_3 - \lambda_2)}\right]$$

$$= N_1(0)\lambda_1\lambda_2\left[\frac{e^{-\lambda_1 t}}{(\lambda_1 - \lambda_2)(\lambda_1 - \lambda_3)} + \frac{e^{-\lambda_2 t}}{(\lambda_2 - \lambda_1)(\lambda_2 - \lambda_3)} + \frac{e^{-\lambda_3 t}}{(\lambda_3 - \lambda_1)(\lambda_3 - \lambda_2)}\right] \quad (6.14)$$

This expression gives the number of atoms of substance 3 at time t. Now, the number of atoms of substance 4 decreases because the substance 4 decays into the substance 5, and increases because of the decay of the substance 3 into the substance 4. Inserting this value of N_3 from equation (6.14) in equation (6.9), we have

$$\frac{dN_4}{dt} + \lambda_4 N_4 = \lambda_1\lambda_2\lambda_3 N_1(0)\left[\frac{e^{-\lambda_1 t}}{(\lambda_1 - \lambda_2)(\lambda_1 - \lambda_3)} + \frac{e^{-\lambda_2 t}}{(\lambda_2 - \lambda_1)(\lambda_2 - \lambda_3)} + \frac{e^{-\lambda_3 t}}{(\lambda_3 - \lambda_1)(\lambda_3 - \lambda_2)}\right]$$

Multiplying this equation throughout by $e^{\lambda_4 t}$ and then integrating, we have

$$N_4 e^{\lambda_4 t} = \lambda_1\lambda_2\lambda_3 N_1(0)\left[\frac{e^{(\lambda_4 - \lambda_1)t}}{(\lambda_1 - \lambda_2)(\lambda_1 - \lambda_3)(\lambda_4 - \lambda_1)} + \frac{e^{(\lambda_4 - \lambda_2)t}}{(\lambda_2 - \lambda_1)(\lambda_2 - \lambda_3)(\lambda_4 - \lambda_2)} + \frac{e^{(\lambda_4 - \lambda_3)t}}{(\lambda_3 - \lambda_1)(\lambda_3 - \lambda_2)(\lambda_4 - \lambda_3)}\right] + C''' \quad (6.15)$$

where C''' is the constant of integration. In the beginning, only the parent substance 1 was present. Thus, at $t = 0$, we have $N_4 = 0$ and therefore from equation (6.15), we get

$$C''' = -\lambda_1\lambda_2\lambda_3 N_1(0)\left[\frac{1}{(\lambda_1 - \lambda_2)(\lambda_1 - \lambda_3)(\lambda_4 - \lambda_1)}\right.$$

$$+\frac{1}{(\lambda_2-\lambda_1)(\lambda_2-\lambda_3)(\lambda_4-\lambda_2)}+\frac{1}{(\lambda_3-\lambda_1)(\lambda_3-\lambda_2)(\lambda_4-\lambda_3)}\Big]$$

Substituting the value of C''' in equation (6.15) and multiplying throughout by $e^{-\lambda_4 t}$, we get

$$N_4=\lambda_1\lambda_2\lambda_3N_1(0)\Big[\frac{e^{-\lambda_1 t}-e^{-\lambda_4 t}}{(\lambda_1-\lambda_2)(\lambda_1-\lambda_3)(\lambda_4-\lambda_1)}$$

$$+\frac{e^{-\lambda_2 t}-e^{-\lambda_4 t}}{(\lambda_2-\lambda_1)(\lambda_2-\lambda_3)(\lambda_4-\lambda_2)}+\frac{e^{-\lambda_3 t}-e^{-\lambda_4 t}}{(\lambda_3-\lambda_1)(\lambda_3-\lambda_2)(\lambda_4-\lambda_3)}\Big]$$

$$=\lambda_1\lambda_2\lambda_3N_1(0)\Big[\frac{e^{-\lambda_1 t}}{(\lambda_1-\lambda_2)(\lambda_1-\lambda_3)(\lambda_4-\lambda_1)}$$

$$+\frac{e^{-\lambda_2 t}}{(\lambda_2-\lambda_1)(\lambda_2-\lambda_3)(\lambda_4-\lambda_2)}+\frac{e^{-\lambda_3 t}}{(\lambda_3-\lambda_1)(\lambda_3-\lambda_2)(\lambda_4-\lambda_3)}$$

$$-e^{-\lambda_4 t}\Big\{\frac{1}{(\lambda_1-\lambda_2)(\lambda_1-\lambda_3)(\lambda_4-\lambda_1)}+\frac{1}{(\lambda_2-\lambda_1)(\lambda_2-\lambda_3)(\lambda_4-\lambda_2)}$$

$$+\frac{1}{(\lambda_3-\lambda_1)(\lambda_3-\lambda_2)(\lambda_4-\lambda_3)}\Big\}\Big]$$

The value of the terms in the curly brackets is[3]

$$\frac{1}{(\lambda_4-\lambda_1)(\lambda_4-\lambda_2)(\lambda_4-\lambda_3)}$$

---

[3]We have

$$\Big\{\frac{1}{(\lambda_1-\lambda_2)(\lambda_1-\lambda_3)(\lambda_4-\lambda_1)}+\frac{1}{(\lambda_2-\lambda_1)(\lambda_2-\lambda_3)(\lambda_4-\lambda_2)}$$

$$+\frac{1}{(\lambda_3-\lambda_1)(\lambda_3-\lambda_2)(\lambda_4-\lambda_3)}\Big\}$$

$$=\frac{A+B+C}{(\lambda_1-\lambda_2)(\lambda_1-\lambda_3)(\lambda_2-\lambda_3)(\lambda_4-\lambda_1)(\lambda_4-\lambda_2)(\lambda_4-\lambda_3)} \tag{6.16}$$

where

$$A=(\lambda_2-\lambda_3)(\lambda_4-\lambda_2)(\lambda_4-\lambda_3)=\lambda_2\lambda_4^2-\lambda_2^2\lambda_4-\lambda_3\lambda_4^2+\lambda_2^2\lambda_3+\lambda_3^2\lambda_4-\lambda_2\lambda_3^2 \tag{6.17}$$

$$B=(\lambda_3-\lambda_1)(\lambda_4-\lambda_1)(\lambda_4-\lambda_3)=\lambda_3\lambda_4^2-\lambda_1\lambda_4^2+\lambda_1^2\lambda_4-\lambda_3^2\lambda_4+\lambda_1\lambda_3^2-\lambda_1^2\lambda_3 \tag{6.18}$$

and

$$C=(\lambda_1-\lambda_2)(\lambda_4-\lambda_1)(\lambda_4-\lambda_2)=\lambda_1\lambda_4^2-\lambda_1^2\lambda_4-\lambda_2\lambda_4^2+\lambda_1^2\lambda_2+\lambda_2^2\lambda_4-\lambda_1\lambda_2^2 \tag{6.19}$$

Substituting the values of equations (6.17), (6.18) and (6.19) in equation (6.16), we get

$$\frac{1}{(\lambda_4-\lambda_1)(\lambda_4-\lambda_2)(\lambda_4-\lambda_3)}$$

Therefore, we have

$$N_4 = \lambda_1\lambda_2\lambda_3 N_1(0)\left[\frac{e^{-\lambda_1 t}}{(\lambda_1-\lambda_2)(\lambda_1-\lambda_3)(\lambda_4-\lambda_1)} + \frac{e^{-\lambda_2 t}}{(\lambda_2-\lambda_1)(\lambda_2-\lambda_3)(\lambda_4-\lambda_2)} + \frac{e^{-\lambda_3 t}}{(\lambda_3-\lambda_1)(\lambda_3-\lambda_2)(\lambda_4-\lambda_3)} - \frac{e^{-\lambda_4 t}}{(\lambda_4-\lambda_1)(\lambda_4-\lambda_2)(\lambda_4-\lambda_3)}\right]$$

This expression gives the number of atoms of substance 4 at time t. Though the series may proceed further, but we assume that the substance 4 is not decaying further. Thus, number of atoms of substance 4 will go on increasing continuously. Ultimately, we shall have only atoms of substance 4.

Exercise 9: A radioactive substance 1 decays into the substance 2 with radioactive constant λ_1. The substance 2 is also radioactive and it decays into the substance 3 with radioactive constant λ_2. The substance 3 is not radioactive. Derive the expressions for the amount of the substances 1, 2 and 3 as a function of time.

Solution: Suppose at time $t = 0$, we have $N_1(0)$ atoms of the radioactive substance 1. Now, we want to find the number of atoms, $N_1(t)$, $N_2(t)$ and $N_3(t)$ present at any subsequent time t. The decay constants for the nuclei are λ_1 and λ_2, respectively. Obviously, the product nuclei $N_2(t)$ and $N_3(t)$ are zero at $t = 0$.

The differential equations representing the number of atoms of each member of the series can be expressed as

$$\frac{dN_1}{dt} = -\lambda_1 N_1 \tag{6.20}$$

$$\frac{dN_2}{dt} = \lambda_1 N_1 - \lambda_2 N_2 \tag{6.21}$$

$$\frac{dN_3}{dt} = \lambda_2 N_2 \tag{6.22}$$

Solution of the differential equation (6.20) can be expressed as

$$N_1 = N_1(0)\ e^{-\lambda_1 t} \tag{6.23}$$

where $N_1(0)$ is the number of atoms of substance 1 at $t = 0$. This expression (6.23) gives the number of atoms of substance 1 at time t.

Now, the number of atoms of substance 2, increases because of the decay of the substance 1 into the substance 2, and decreases because of the decay of the substance 2 into the substance 3. Inserting this value of N_1 from equation (6.23) in (6.21) and rearranging, we have

$$\frac{dN_2}{dt} + \lambda_2 N_2 = \lambda_1 N_1(0)\ e^{-\lambda_1 t}$$

Multiplying this equation throughout by $e^{\lambda_2 t}$ and then integrating, we have

$$N_2\ e^{\lambda_2 t} = \frac{\lambda_1 N_1(0)}{\lambda_2 - \lambda_1}\ e^{(\lambda_2 - \lambda_1)t} + C' \tag{6.24}$$

where C' is the constant of integration. Since in the beginning, only the parent substance 1 was present. Thus, at $t = 0$, we have $N_2 = 0$ and therefore from equation (6.24), we get

$$C' = -\frac{\lambda_1 N_1(0)}{\lambda_2 - \lambda_1}$$

Substituting value of C' in equation (6.24) and multiplying throughout by $e^{-\lambda_2 t}$, we get

$$N_2 = \frac{\lambda_1 N_1(0)}{(\lambda_2 - \lambda_1)}\left[e^{-\lambda_1 t} - e^{-\lambda_2 t}\right] \tag{6.25}$$

This expression gives the number of atoms of substance 2 at time t. Now, the number of atoms of substance 3 goes on increasing with time as it does not decay and the substance 2 decays into the substance 3. Inserting this value of N_2 from equation (6.25) in (6.22), we have

$$\frac{dN_3}{dt} = \frac{\lambda_1 \lambda_2 N_1(0)}{(\lambda_2 - \lambda_1)}\left[e^{-\lambda_1 t} - e^{-\lambda_2 t}\right]$$

On integrating this equation, we have

$$N_3 = \frac{\lambda_1 \lambda_2 N_1(0)}{(\lambda_2 - \lambda_1)}\left[-\frac{e^{-\lambda_1 t}}{\lambda_1} + \frac{e^{-\lambda_2 t}}{\lambda_2}\right] + C'' \tag{6.26}$$

where C'' is the constant of integration. In the beginning, only the parent substance 1 was present. Thus, at $t = 0$, we have $N_3 = 0$ and therefore from equation (6.26), we get

$$C'' = -\frac{\lambda_1 \lambda_2 N_1(0)}{(\lambda_2 - \lambda_1)}\left[-\frac{1}{\lambda_1} + \frac{1}{\lambda_2}\right]$$

Substituting the value of C'' in equation (6.26), we get[4]

$$N_3 = \frac{\lambda_1 \lambda_2 N_1(0)}{(\lambda_2 - \lambda_1)} \left[\frac{1 - e^{-\lambda_1 t}}{\lambda_1} - \frac{1 - e^{-\lambda_2 t}}{\lambda_2}\right]$$

This expression gives the number of atoms of substance 3 at time t. The variation of N_1, N_2 and N_3 as a function time t for the case that the radioactive constants for the substances 1 and 2 are in the ratio 5 : 3 is shown in Figure 3.

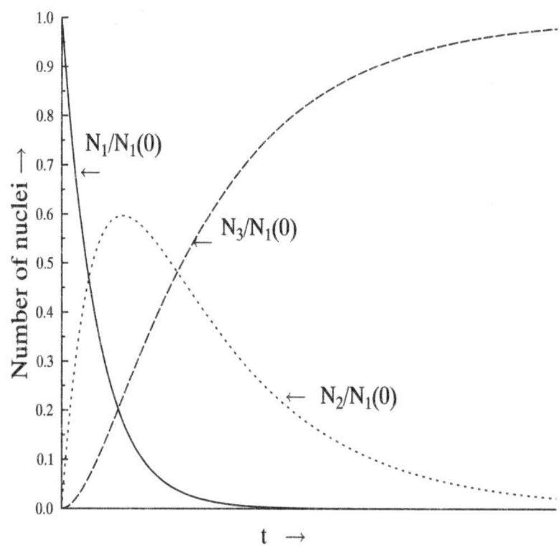

Figure 3: Variation N_1, N_2 and N_3 as a function time t for the case that the radioactive constants for the substances 1 and 2 are in the ratio 5 : 3.

Exercise 10: A radioactive substance 1 decays into the substance 2 with radioactive constant λ. The substance 2 is also radioactive and it decays into the substance 3 with the same radioactive constant λ. The substance 3 is not radioactive. Derive the expressions for the amount of the substances 1, 2 and 3 as a function of time.

---

[4]We have

$$N_3 = \frac{\lambda_1 \lambda_2 N_1(0)}{(\lambda_2 - \lambda_1)} \left[\frac{1 - e^{-\lambda_1 t}}{\lambda_1} - \frac{1 - e^{-\lambda_2 t}}{\lambda_2}\right]$$

$$= \frac{\lambda_2 N_1(0)}{(\lambda_2 - \lambda_1)} - \frac{\lambda_2 N_1(0)}{(\lambda_2 - \lambda_1)} e^{-\lambda_1 t} - \frac{\lambda_1 N_1(0)}{(\lambda_2 - \lambda_1)} + \frac{\lambda_1 N_1(0)}{(\lambda_2 - \lambda_1)} e^{-\lambda_2 t}$$

$$= N_1(0) - \frac{\lambda_2 N_1(0)}{(\lambda_2 - \lambda_1)} e^{-\lambda_1 t} + \frac{\lambda_1 N_1(0)}{(\lambda_2 - \lambda_1)} e^{-\lambda_1 t} - \frac{\lambda_1 N_1(0)}{(\lambda_2 - \lambda_1)} e^{-\lambda_1 t} + \frac{\lambda_1 N_1(0)}{(\lambda_2 - \lambda_1)} e^{-\lambda_2 t}$$

$$= N_1(0) - N_1 - N_2$$

Solution: Suppose at time $t = 0$, we have $N_1(0)$ atoms of the radioactive substance 1. Now, we want to find the number of atoms, $N_1(t)$, $N_2(t)$ and $N_3(t)$ present at any subsequent time t. The decay constants for the nuclei are the same λ. Obviously, the product nuclei $N_2(t)$ and $N_3(t)$ are zero at $t = 0$. The differential equations representing the number of atoms of each member of the series can be expressed as

$$\frac{dN_1}{dt} = -\lambda N_1 \tag{6.27}$$

$$\frac{dN_2}{dt} = \lambda N_1 - \lambda N_2 \tag{6.28}$$

$$\frac{dN_3}{dt} = \lambda N_2 \tag{6.29}$$

Solution of the differential equation (6.27) can be expressed as

$$N_1 = N_1(0)\ e^{-\lambda t} \tag{6.30}$$

where $N_1(0)$ is the number of atoms of substance 1 at $t = 0$. This expression (6.30) gives the number of atoms of substance 1 at time t. Now, the number of atoms of substance 2, increases because of the decay of the substance 1 into the substance 2, and decreases because of the decay of the substance 2 into the substance 3. Inserting this value of N_1 from equation (6.30) in (6.28) and rearranging, we have

$$\frac{dN_2}{dt} + \lambda N_2 = \lambda N_1(0)\ e^{-\lambda t}$$

Multiplying this equation throughout by $e^{\lambda t}$ and then integrating, we have

$$N_2\ e^{\lambda t} = \lambda N_1(0)t + C' \tag{6.31}$$

where C' is the constant of integration. Since in the beginning, only the parent substance 1 was present. Thus, at $t = 0$, we have $N_2 = 0$ and therefore, we get $C' = 0$. Substituting value of C' in equation (6.31) and multiplying throughout by $e^{-\lambda t}$, we get

$$N_2 = \lambda N_1(0)t\ e^{-\lambda t} \tag{6.32}$$

This expression gives the number of atoms of substance 2 at time t. Now, the number of atoms of substance 3 goes on increasing with time as it does not decay and the substance 2 decays into the substance 3. Inserting this value of N_2 from equation (6.32) in (6.29), we have

$$\frac{dN_3}{dt} = \lambda^2 N_1(0)t\ e^{-\lambda t}$$

On integrating this equation, we have

$$N_3 = N_1(0)[-t\lambda\, e^{-\lambda t} - e^{-\lambda t}] + C'' \tag{6.33}$$

where C'' is the constant of integration. In the beginning, only the parent substance 1 was present. Thus, at $t = 0$, we have $N_3 = 0$ and therefore from equation (6.33), we get $C'' = N_1(0)$. Substituting the value of C'' in equation (6.33), we get[5]

$$N_3 = N_1(0)[1 - e^{-\lambda t}(1 + t\lambda)]$$

This expression gives the number of atoms of substance 3 at time t. The variation of N_1, N_2 and N_3 as a function time t is shown in Figure 4.

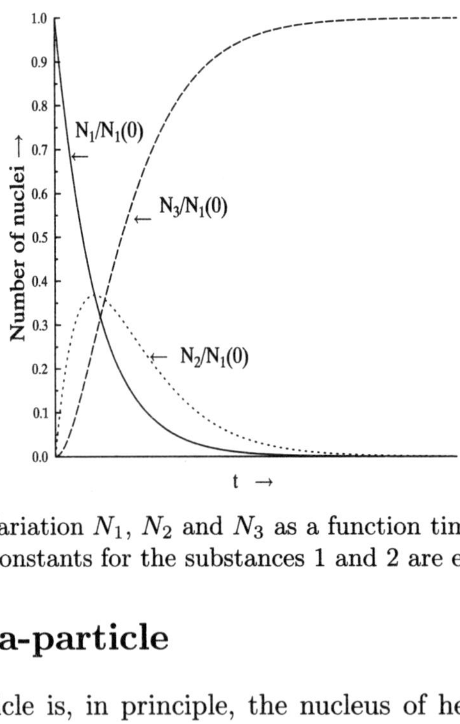

Figure 4: Variation N_1, N_2 and N_3 as a function time t for the case that the radioactive constants for the substances 1 and 2 are equal.

3. Alpha-particle

Alpha particle is, in principle, the nucleus of helium atom. It carries mass 4 and charge +2. It is one of the particles emitted during disintegration of radioactive element. Let us investigate various aspects of the α-particle.

3.1 Energy spectrum of α-particles

In the beginning, it was thought that all the α-particles emitted from a large group of parent nuclei would have the same energy. However,

[5] We can easily show that $N_3 = N_1(0) - N_1 - N_2$

accurate measurements taken with the help of magnetic spectrograph showed that the α-particles had different energies falling within a narrow range. Moreover, the energies of the particles have been found to lie in various groups. For example, ThC nucleus is found to emit α-particles lying in five groups with discrete energies of 5.709, 5.728, 5.873, 6.161 and 6.201 MeV. The emission of groups of α-particles with different energies suggests that nuclei may exit in various discrete excited states above the ground state. This feature is known as the energy spectrum of α-particles.

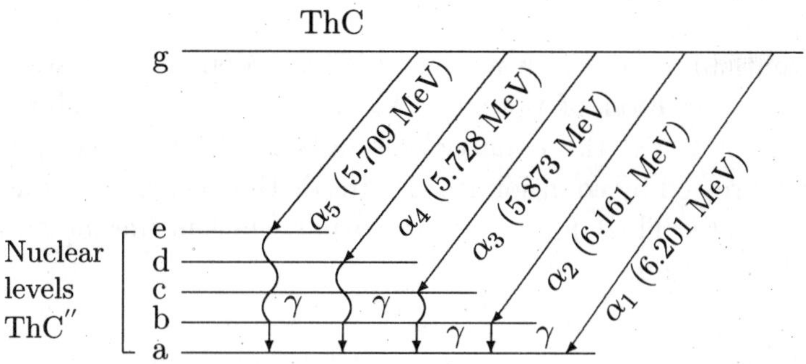

Figure 5: The parent nuclei ThC in the ground state g emit α-particle. The daughter nuclei ThC$''$ are formed in various excited states b, c, d, e, or in the ground state a.

It can be understood with the help of Figure 5, where the parent nuclei ThC are in the ground state g. After emission of α-particle, the daughter nuclei (ThC$''$) may exit in any of the excited states b, c, d, e, or in the ground state a. Thus, the energy of the emitted α-particle will depend on the state of the daughter nucleus. The α-particle will have the largest energy when the formed daughter nucleus is in the ground state.

Later on, the daughter nucleus formed in the excited state, deexcites to the ground state a, and a γ-radiation with the corresponding energy is produced.

3.2 Theory of α-particle emission

A heavy nucleus with mass number $A > 200$ undergoes the emission of α-particle. It is also known as the α-decay. Energy of these emitted α-particles show a spectrum, *i.e.*, the emitted α-particles are found to have various energies lying in a narrow range. No α-decay is allowed classically. However, the quantum mechanics, in 1928, explained successfully the phenomenon of α-decay through the tunneling effect. Before emission, the α-particle is inside the parent nucleus. When the α-particle comes outside the nucleus, there is Coulomb repulsion between the α-particle and the daughter nucleus. Suppose Z be the atomic number of the parent nucleus. After emission of α-particle, the atomic number of the daughter nucleus is $(Z - 2)$. The potential energy between the α-particle and the daughter nucleus is expressed as

$$V_r = \frac{2e(Z-2)e}{r} = \frac{2(Z-2)e^2}{r} \tag{6.34}$$

This potential is spherically symmetric, *i.e.*, it depends on the radial distance only. Variation of potential V_r with distance r is shown in Figure 6. For $r > r_0$, the α-particle is outside the nucleus and the potential is expressed by equation (6.34). Inside the nucleus, the nuclear forces predominate. Thus, for $r < r_0$, the α-particle is inside the nucleus and it is tightly bound.

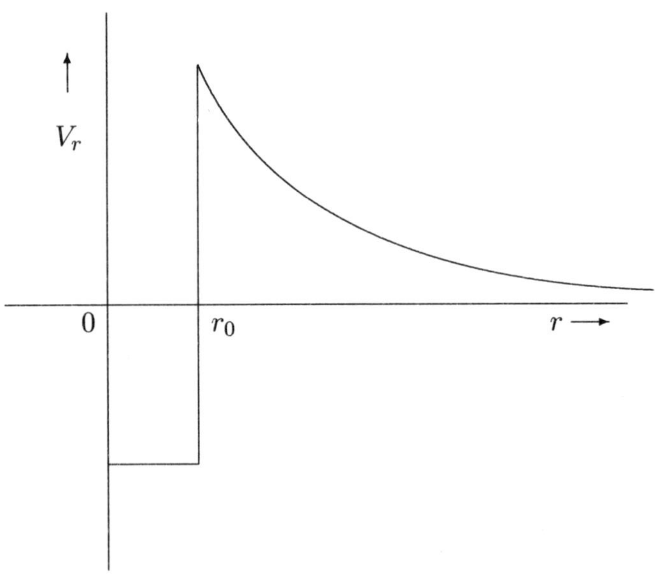

Figure 6: Variation of potential V_r with distance r. For $r > r_0$, the α-particle is outside the nucleus.

It is the case of potential barrier in quantum mechanics. Quantum mechanics supports the idea that even the energy of a particle is less than the height of the barrier, there is a finite probability for the transmission of particle through the barrier. It is known as the tunneling effect. Thus, the quantum mechanics explains the process of α-decay.

3.3 Gamow Theory for α-decay

Classical mechanics does not allow the phenomenon of α-decay. However, the quantum mechanics, in 1928, explained successfully the phenomenon of α-decay through the tunneling effect. The basic notions of this theory are as the following.

(i) An α-particle may exist as an entity within a nucleus

(ii) The α-particle is in motion and bounces back and forth after colliding at the barrier walls (Figure 6). In each collision with the wall, there is definite probability that the particle will transmit through the barrier. Let ν be the frequency of collision of α-particle with the wall and P the probability of transmission of α-particle through the barrier. Then the decay probability per unit time (*i.e.*, the disintegration constant λ) is

$$\lambda = \nu P$$

It is spherically symmetric case, *i.e.*, it depends on the radial distance only. The variation of potential $V(r)$ with distance r is shown in Figure 6.

Here, the height of the potential barrier is larger than the kinetic energy E of the particle. According to quantum mechanics, the probability P for the transmission through a potential barrier is expressed as

$$\ln P = -\frac{2}{\hbar}\int_{r_0}^{r_1} \sqrt{2m[V(r) - E]}\, dr \qquad (6.35)$$

where m is the mass of α-particle and the potential energy $V(r)$ can be approximated as

$$V(r) = \frac{2Ze^2}{r}$$

The region from $r = r_0$ to $r = r_1$ is known as the thickness of the barrier. At $r = r_1$, $V = E$ and therefore we have

$$E = \frac{2Ze^2}{r_1}$$

Using the expression for $V(r)$ and E in equation (6.35), we get

$$\ln P = -\frac{2\sqrt{2mE}}{\hbar}\int_{r_0}^{r_1} \left[\frac{r_1}{r} - 1\right]^{1/2} dr$$

$$= -\frac{2r_1\sqrt{2mE}}{\hbar}\left[\cos^{-1}\left(\frac{r_0}{r_1}\right)^{1/2} - \left(\frac{r_0}{r_1}\right)^{1/2}\left(1-\frac{r_0}{r_1}\right)^{1/2}\right] \quad (6.36)$$

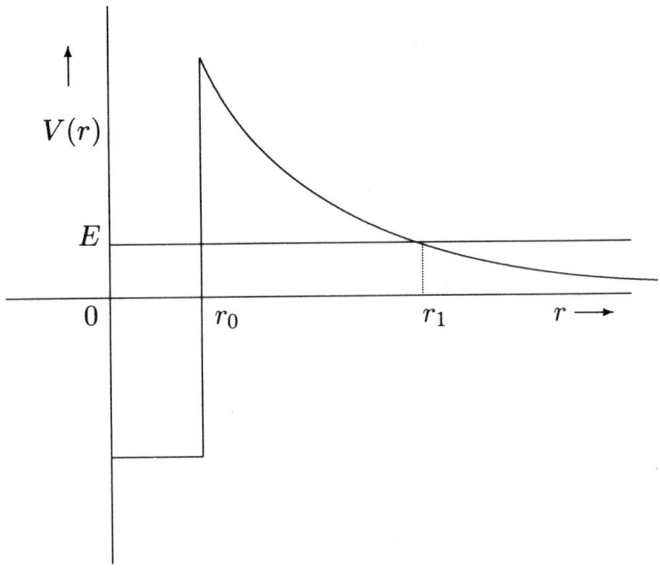

Figure 7: Variation of potential $V(r)$ with distance r. For $r > r_0$, the α-particle is outside the nucleus.

The width of the potential barrier is very large as compared to the radius of nucleus, *i.e.*, $r_1 >> r_0$ and therefore[6]

$$\cos^{-1}\left(\frac{r_0}{r_1}\right)^{1/2} = \frac{\pi}{2} - \left(\frac{r_0}{r_1}\right)^{1/2} \qquad \text{and} \qquad \left(1-\frac{r_0}{r_1}\right)^{1/2} = 1$$

Thus, equation (6.36) reduces to

$$\ln P = -\frac{2r_1\sqrt{2mE}}{\hbar}\left[\frac{\pi}{2} - 2\left(\frac{r_0}{r_1}\right)^{1/2}\right] = \frac{\sqrt{2mE}}{\hbar}\left[4r_0^{1/2}r_1^{1/2} - \pi r_1\right] \quad (6.37)$$

Using $r_1 = 2Ze^2/E$ in equation (6.37), we get

$$\ln P = \frac{8e\sqrt{m}}{\hbar}\, Z^{1/2}r_0^{1/2} - \frac{\pi 2\sqrt{2m}Ze^2}{\hbar}\, E^{-1/2}$$

[6]We know that

$$y = \cos^{-1} x \qquad x = \cos y \qquad x = \pi/2 - \sin y$$

When y is very small, we have $\sin y = y$ and therefore

$$x = \pi/2 - y \qquad y = \pi/2 - x$$

Using the values of constants, we get

$$\ln P = 2.97Z^{1/2}r_0^{1/2} - 3.95ZE^{-1/2} \tag{6.38}$$

where r_0 is in Fermi and E in MeV. Now, we have

$$\lambda = \nu P \qquad \text{thus} \qquad \ln \lambda = \ln \nu + \ln P$$

Using equation (6.38), we get

$$\ln \lambda = \ln \nu + 2.97Z^{1/2}r_0^{1/2} - 3.95ZE^{-1/2}$$

Changing the base of logarithm from e to 10, we have[7]

$$\log \lambda = \log \nu + 1.29Z^{1/2}r_0^{1/2} - 1.72ZE^{-1/2} \tag{6.39}$$

The changes in atomic number and nuclear radius are negligible as compared to the changes in energy. Thus, equation (6.39) can be expressed as

$$\log \lambda = c + d\ E^{-1/2} \tag{6.40}$$

where c and d are constants. Equation (6.40) shows that the emitter having smaller decay constant emits α-particle of larger energy. It is known as the Geiger and Nuttal law.

3.4 Velocity and energy

For determination of velocity and energy of α-particles, one should pay specific attention for several reasons. Some reasons may be as the following.

(i) The accurate measurement of energy of α-particles, it would be possible to determine energies which differ only by small amounts. This in fact led to the discovery that some radioactive nuclei actually emit a spectrum of α-particles.

(ii) Knowledge of the energies of the components of alpha spectra makes it possible to assign certain nuclear energy levels with confidence.

---

[7] We know

$$\ln x = \log x \ln 10 \qquad \text{therefore} \qquad \log x = \ln x \times 0.4343$$

(iii) The methods for determination of energies of α-particles can also be used for protons and deuterons. These three charged particles are involved in many artificial nuclear reactions. Obviously, the accurate measurement of their energies would help in getting accurate Q-values for the reactions. For the known Q-values, it is possible to determine nuclear masses and nuclear energy levels.

(iv) Accurate values of energies for α-particles are required in the development and for the use in the theory of α-decay.

The method which gives the most accurate results for the velocity and energy of α-particles depends on the measurement of deflection of the path of the particles in a magnetic field. When the direction of the applied magnetic field is perpendicular to the direction of motion of the charged particle, its orbit is circle whose radius r is determined by the relation

$$qvB = \frac{mv^2}{r} \tag{6.41}$$

where B is the strength of the magnetic field applied, q and m are the charge and mass of the particle, respectively. Equation (6.41) may be rearranged as

$$v = \frac{q}{m}\,Br$$

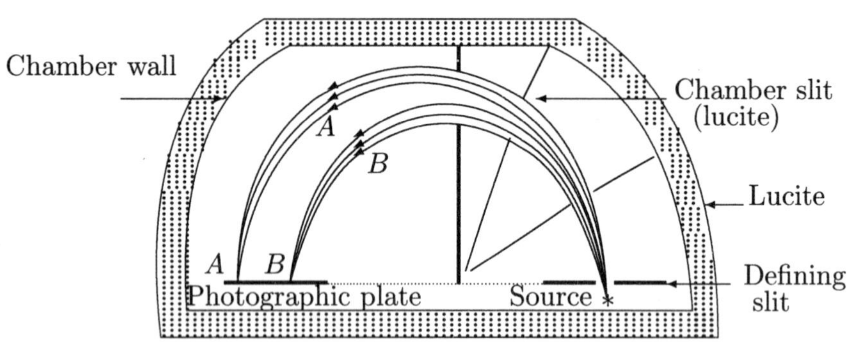

Figure 8: Schematic diagram for deflection chamber of a magnetic spectrograph for α-particles

For the α-particles, the charge to mass ratio is known. For the given magnetic field strength, by measuring the radius of the orbit, the velocity of the α-particle can be determined.

Figure 8 is the schematic diagram of an apparatus used for the purpose. This apparatus, called the magnetic spectrograph, is based on the said above principle. A narrow beam of α-particles, from a radioactive source, moving in the plane of the diagram is produced with the help of a slit. A magnetic field of known strength acting in the direction perpendicular to the plane of the diagram is applied. This field bends the α-particles through an angle of 180°. The slits in the chamber help to reduce the scattering of α-particles from the top, bottom and walls of the chamber. Particles with the same velocity have semicircular paths of the same radius. They may be detected with the help of a photographic plate or a counter and the radius of the path can be measured. Figure 8 shows paths of two groups of particles with different velocities. The α-particles can be bent into a semicircle of radius of 40 or 50 cm maximum.

For small velocities, the relativistic mass correction is small and is often neglected. In this case, the kinetic energy E of the particle is $mv^2/2$. When the velocity is large enough so that the relativity correction must be accounted for, then the velocity v is expressed as

$$v = Br\frac{q}{m_0}\sqrt{1 - \frac{v^2}{c^2}} \tag{6.42}$$

and the relativistic kinetic energy as

$$T = m_0c^2\left[\frac{1}{\sqrt{1 - (v^2/c^2)}} - 1\right] \tag{6.43}$$

For the applied magnetic field B and the measured radius r, the velocity is obtained from equation (6.42). For the known velocity v, the kinetic energy is obtained from equation (6.43).

3.5 Range

Bragg was the first to suggest that when an α-particle passes through a gas, it looses energy for ionization of gas particles until its energy falls below the ionization potential of the gas. After that, the captures two electrons to become a neutral helium atom. The length of the path traveled by the particle till it loses its ionization power is known as the range of the α-particle. This range obviously depends on the initial energy of the particle, the ionization potential of the gas, and the nature and pressure of the gas. The range will be considerably low in solids and liquids where the material particles are densely packed.

3.6 α-disintegration energy

The process in which an α-particle is spontaneously ejected from a nucleus is known as the α-decay. Emission of α-particle from a nucleus was first successfully explained on the basis of the tunneling effect in quantum mechanics 1928. In the emission of an α-particle, the parent nucleus looses an aggregate of two protons and two neutrons. The process may be represented as

$$ {}_Z X^A \rightarrow {}_2 He^4 + {}_{Z-2} Y^{A-4} + Q_\alpha $$

where ${}_Z X^A$ represents the parent nucleus and ${}_{Z-2}Y^{A-4}$ the daughter nucleus. The quantity Q_α is known as the α-disintegration energy and represents the total energy released in the process. This energy consists of the kinetic energy of the α-particle emitted and the kinetic energy of the product nucleus. The value of Q_α corresponds to the difference in mass of the parent nucleus and the product nucleus.

For the evaluation of the value of Q_α, let us consider that the parent nucleus of mass m_1 is at rest. Suppose the nucleus emits an α-particle of mass m moving with velocity v, the residual atom of mass m_2 will recoil with a velocity v_2 , such as

$$ m_2 v_2 = mv \tag{6.44} $$

The disintegration energy is

$$ Q_\alpha = \frac{1}{2}\, m_2 v_2^2 + \frac{1}{2}\, mv^2 \tag{6.45} $$

On eliminating v_2 from equations (6.44) and (6.45), we get

$$ Q_\alpha = W_\alpha \left(\frac{m}{m_2} + 1 \right) $$

where $W_\alpha = mv^2/2$ is the kinetic energy of α-particle emitted. In the first approximation, we can replace the ratio of the masses with the ratio of the mass-numbers, as

$$ \frac{m}{m_2} = \frac{4}{A-4} $$

where A is the mass number of the parent atom. Thus, we have

$$ Q_\alpha = \frac{A}{A-4} W_\alpha \qquad \text{or} \qquad W_\alpha = \left(1 - \frac{4}{A}\right) Q_\alpha $$

This equation shows that most of the kinetic energy released during the α-decay is carried away by the α-particle and a small fraction remains with the daughter nucleus.

Exercise 11: A stationary nucleus with atomic mass $A = 100$ disintegrates through emission of α-particle. Calculate the ratio of the kinetic energies of the emitted α-particle and the daughter nucleus.

Solution: For the process

$$X^{100} \rightarrow He^4 + Y^{96} + Q_\alpha$$

the energy released Q_α is shared by the α-particle and the daughter nucleus. The kinetic energy W_α of the α-particle is

$$W_\alpha = \left(1 - \frac{4}{A}\right)Q_\alpha = \left(1 - \frac{4}{100}\right)Q_\alpha = 0.96Q_\alpha$$

The kinetic energy W_Y of the daughter nucleus Y is

$$W_Y = Q_\alpha - 0.96Q_\alpha = 0.04Q_\alpha$$

The ratio of the kinetic energies is

$$\frac{W_\alpha}{W_Y} = \frac{0.96Q_\alpha}{0.04Q_\alpha} = 24$$

Exercise 12: Polonium-212 emits α-particle having kinetic energy of 10.54 MeV. Calculate the α-disintegration energy.

Solution: After emission of α-particle from Polonium-212, the mass of the product nucleus is $m_2 = 208$. The α-disintegration energy is

$$Q_\alpha = W_\alpha\left(1 + \frac{m}{m_2}\right) = 10.54\left(1 + \frac{4}{208}\right) = 10.74 \text{ MeV}$$

4. Beta particle

Beta particle is, in principle, the electron. It carries charge -1 and negligible mass as compared to the mass of proton or neutron. It is one of the particles emitted during disintegration of radioactive element. Let us investigate various aspects of the β-particle.

4.1 Energy spectrum of β-particles

In a given magnetic field B, a β-particle traverses a circular path in a plane whose normal is along the field direction. The radius of curvature r of the path depends on the velocity v and is expressed as

$$evB = \frac{m_e v^2}{r} \qquad \text{or} \qquad v = Br(e/m_e) \tag{6.46}$$

where e and m_e are charge and mass of electron (β-particle). For the β-particles with different velocities (energies), the radii of curvature are different. Since the magnetic field B and the ratio (e/m_e) are known, for the measured value of r, we can calculate the velocity (energy) for various β-particles. This equation (6.46) is applicable for the non-relativistic case, *i.e.*, the velocity $v << c$.

When the velocity of a β-particle is comparable to the speed of light, relativistic corrections come in the picture and the kinetic energy E_k is expressed as

$$E_k = mc^2 - m_0c^2 = m_0c^2\Big[\frac{1}{\sqrt{1 - v^2/c^2}} - 1\Big]$$

Emission of β-particles during the radioactive decay of a parent nucleus is known as the energy spectrum of β-particles.

Exercise 13: Kinetic energy of a β-particle is 0.5 MeV. Calculate the velocity and mass of the particle. When a beam of such particles enters a uniform magnetic field, calculate the magnetic field strength required to cause them to traverse a circular arc of radius 10 cm.

Solution: The relativistic kinetic energy of β-particle is

$$E_k = m_0c^2\Big[\frac{1}{\sqrt{1 - v^2/c^2}} - 1\Big]$$

The kinetic energy $E_k = 0.5$ MeV $= 0.5 \times 10^6 \times 1.602 \times 10^{-19} = 8.01 \times 10^{-14}$ J. Using $m_0 = 9.11 \times 10^{-31}$ kg and $c = 3 \times 10^8$ m/s, in the expression for the relativistic kinetic energy, we have

$$8.01 \times 10^{-14} = 9.11 \times 10^{-31} \times (3 \times 10^8)^2\Big[\frac{1}{\sqrt{1 - v^2/(3 \times 10^8)^2}} - 1\Big]$$

$$\Big[\frac{1}{\sqrt{1 - v^2/(3 \times 10^8)^2}} - 1\Big] = 0.976948$$

It gives $v = 2.59 \times 10^8$ m/s $= 0.863\ c$. Mass of β-particle is

$$m = \frac{m_0}{\sqrt{1 - v^2/c^2}} = \frac{9.11 \times 10^{-31}}{\sqrt{1 - 0.863^2}} = 1.8 \times 10^{-30} \text{ kg}$$

Thus, the mass of β-particle is 1.98 m_0. The required magnetic field strength B is

$$B = \frac{mv}{er} = \frac{1.8 \times 10^{-30} \times 2.59 \times 10^8}{1.6 \times 10^{-190} \times 0.1} = 0.02914 \text{ weber/m}^2$$

4.2 Neutrino theory of β-decay

In order to explain the continuous energy spectrum of β-particles, Fermi in 1934, proposed a theory, called the neutrino theory of β-particles. He discussed that both the β-particle and neutrino are created in a nucleus and both of them are emitted from the nucleus simultaneously. Moreover, the sum of the total energy of β-particle and neutrino is conserved. This value of the energy is equal to the end-point energy observed in the spectrum. Obviously, when the energy of the neutrino is zero, the β-particle will have the maximum energy. With the increase of energy of neutrino, the energy of β-particle will decrease. This total amount of energy may be shared between the β-particle and the neutrino in any proportion. This explains the continuous energy spectrum of β-particles.

When in a nucleus, a neutron transforms into a proton, a β-particle and a neutrino are emitted. This process may be represented as

$$n \rightarrow p + \beta^- + \overline{\nu}$$

In an ordinary β-decay, it is an anti-neutrino $\overline{\nu}$ which is emitted. Electron emission leads to a daughter nucleus of higher atomic number Z while leaving the mass number A unchanged.

Positron (β^+) emission corresponds to the transformation of a proton into a neutron. This process may be represented as

$$p \rightarrow n + \beta^+ + \nu$$

Positron emission leads to a daughter nucleus of lower atomic number Z while leaving the mass number A unchanged. Thus, the negative and positive β-decays may be represented as

$$_ZX^A \rightarrow \; _{Z+1}Y^A + \beta^- + \overline{\nu} \qquad \text{and} \qquad _ZX^A \rightarrow \; _{Z-1}Y^A + \beta^+ + \nu$$

The β-particle, neutrino and product nucleus share the energy, momentum and angular momentum available from the nuclear transition. Hence, the neutrino theory of β-decay explains successfully the continuous energy spectrum of β-particles.

Exercise 14: Which of the following would be β^- active and how it would decay. Given, $_{28}\text{Ni}^{64} = 63.9280$ amu and $_{29}\text{Cu}^{64} = 63.9298$ amu.

Solution: We know that a β^- decay is possible when the mass of the parent nucleus $_ZX^A$ is larger than the mass of ($_{Z+1}Y^A$ + β^-). After adding masses of Z electrons on both sides, The nuclear mass change to isotopic masses. Thus, for the β^- decay, the mass of $_ZX^A$ atom is larger than that of $_{Z+1}Y^A$ atom. Since the mass of $_{29}\mathrm{Cu}^{64}$ is larger than that of $_{28}\mathrm{Ni}^{64}$, therefore the possible reaction is

$$_{29}\mathrm{Cu}^{64} \rightarrow \ _{28}\mathrm{Ni}^{64} + \beta^-$$

Thus, $_{29}\mathrm{Cu}^{64}$ is a β^- emitter.

5. Problems and questions

1. Show that during the α-decay, most of the kinetic energy liberated is carried out by the α-particle.

2. A radioactive nucleus A decays in sequence to the radioactive nuclei B, C, D. That is A decays to B, B decays to C, C decays to D and D is a stable nucleus. Suppose the decay constants are λ_1, λ_2, λ_3, respectively, and at $t = 0$, we have $N_1(0)$ atoms of the nucleus A. Derive expressions for the number of atoms $N_1(t)$, $N_2(t)$, $N_3(t)$ and $N_4(t)$ of the nuclei A, B, C and D at time t.

3. A radioactive substance 1 decays into the substance 2 with radioactive constant λ. The substance 2 is also radioactive and it decays into the substance 3 with the same radioactive constant λ. The substance 3 is not radioactive. Derive the expressions for the amount of the substances 1, 2 and 3 as a function of time.

4. Describe the Gamow Theory for α-decay.

5. Write notes on the following

   (i) Natural radioactivity
   (ii) Alpha, beta and gamma rays
   (iii) Properties of alpha, beta and gamma rays
   (iv) Energy spectrum of α-particles
   (v) Energy spectrum of β-particles
   (vi) Gamow Theory for α-decay
   (vii) Neutrino theory of β-decay

VII. Detectors

In most of the nuclear reactions, there is emission of charged particles, such as proton, electron, α-particle, etc. and γ-radiation. For the study of nuclear reactions, one needs to detect and identify the charged as well as neutral particles participating in the given reaction. This can be done in terms of getting information about their mass, momentum, energy, position, etc. Nuclear detectors are the instruments developed for this purpose. Working of these instruments is based on the principle of converting energy of the incident particle/radiation inside the detection medium, producing an output accessible to human perception. The incident radiation produces some kind of charge carriers, which in turn produce some electrical pulses carrying the information about the incident particles. In the present chapter, we shall discuss about different types of nuclear detectors, which can be classified as the gas detectors, ionization chamber, proportional counter, Geiger Müller counter, semiconductor counter and photomultiplier based scintillation detectors. These are the detectors which are commonly used for the detection of particles of various types. Before discussing about detectors, let us first look into some general characteristics of the detectors.

1. General properties of detectors

Before talking about the detectors, let us first discuss about some general characteristics: (i) Sensitivity, (ii) Response, (iii) Resolution, (iv) Efficiency and (v) Dead time

1.1 Sensitivity of a detector

The requirement from a detector is to produce a usable signal corresponding to a given radiation/energy. The use of a specific type of detector depends on the kind of radiation and amount of energy. The sensitivity of a detector defines its capability to generate signals corresponding to certain radiation and energy and it depends on the cross section of the ionizing reaction, detector mass, inherent noise, etc. Obviously, there should be a minimum amount of ionization to generate a usable signal. This lower limit is determined from the noise produced from the detector and the electronics associated with it. Hence, the ionization signal should be higher than the noise of the detector. When the probability of ionization inside the active region of the detector is high, a small volume or active mass of the detector is enough for detection. On the other side, when the probability of reaction is too low, then a larger volume is required.

1.2 Response of a detector

In a detector, the ionization produced is proportional to the loss of energy of radiation in the sensitive volume. When the radiation is absorbed completely, the ionization produced is a measure of energy of the radiation. The output signal of the detector is generally in the form of a current pulse. The charge in this pulse is basically a measure of the amount of ionization. Thus, the radiation energy and the total charge or pulse height of the output signal define the response of the detector.

1.3 Energy resolution of a detector

When a mono-energetic radiation is sent to a detector, one would expect an output signal in the form of delta function with a sharp peak. However, in practice, a Gaussian like distribution with some finite width is obtained. This width is mainly due to statistical fluctuations in the number of ionization and excitation produced. Besides that the intrinsic noise of the system and the errors of the instruments associated with the detector are also responsible for the broadening in the output. All these factors together decide about the resolution of the detection system, *i.e.*, the extent to which the detector can distinguish between two close lying energies. For mono-energetic radiation, resolution is generally expressed in terms of the 'full width at half-maximum' (FWHM) of the peak. The FWHM is shown in Figure 1 and is defined as the width of the distribution at a level, which is just half the maximum ordinate of the peak.

Energies which are smaller than this interval are usually considered unresolvable. Resolution R is defined as the ratio of the FWHM and the peak height H_0. Therefore,

$$R = \frac{\text{FWHM}}{H_0}$$

After multiplying by 100 on the right side, the resolution is sometimes expressed in the percentage.

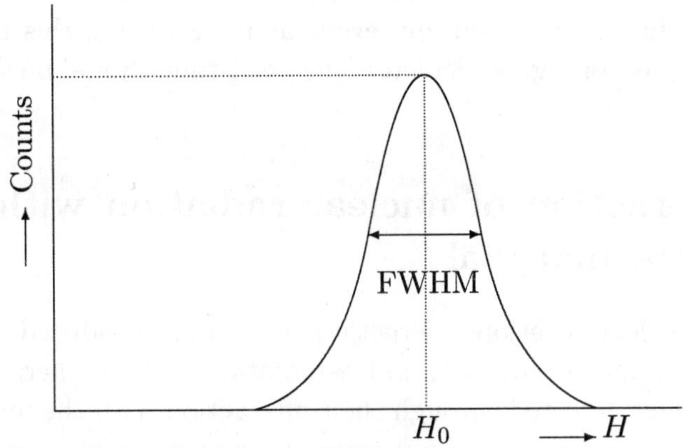

Figure 1: Pulse height versus counts.

1.4 Efficiency of a detector

Any detection system has a specific capability to generate a signal corresponding to the events received from the source. The efficiency ϵ of a detector is expressed as the ratio of the number of events registered by the detector to the number of events produced by the source. Thus, the efficiency is

$$\epsilon = \frac{\text{events registered}}{\text{events emitted by the source}}$$

It depends upon the detector geometry and on the probability of interaction in the detector. There is another way for expressing the efficiency of detector. It is expressed as the ratio of the number of events registered by the detector to the number of events impinging on the detector. Thus, the efficiency is

$$\epsilon = \frac{\text{events registered}}{\text{events impinging on the detector}}$$

The efficiency of a detector depends on the interaction cross section of the incident radiation on the detector medium. Thus, it depends on the type of radiation, its energy, and the detector material.

1.5 Dead time of a detector

When the events take place continuously, it would not be possible to record all of them. For each detector, a minimum interval between two successive events is required for their recording. This minimum time separation between two successive events is generally known as the 'dead time' of the counting system. During the dead time, the detection system is insensitive, and any event arriving during this period is not recorded. So, for a good detector, the dead-time should be the minimum possible.

2. Interaction of nuclear radiation with the detector material

In each nuclear reaction, energetic particles are produced, such as protons, electrons, α-particles, and γ-radiation. These energetic particles are generally detected through their interaction with the material in the detector. Let us review about some mechanisms which provide output for the interaction.

2.1 Heavy charged particles

A charged particle, such as proton, α-particle or fission product has a fairly definite range in a gas, liquid or solid. All the heavy particles carry positive charge. When the charged particle moves through the matter of the detector, it loses its energy for excitation and ionization of atoms in its path. In each interaction, the particle loses energy and there are a large number of interactions, before it comes to rest. The distance traversed by the particle is known as its range.

2.2 Electron

Electron or β-particle interacts with detector material through Coulomb scattering from the atomic electrons in the material, just like the case of heavy charged particles. There are however some differences:

(i) Electrons travel at the relativistic speeds whereas the charged particles move with relativistic speeds.

(ii) Because of low mass, the electrons suffer large deflections in the collisions with other electrons and thus will follow erratic path. Estimation of range will therefore be difficult from the length of the path traversed by the electron.

(iii) Every electron having energy $E > 1$ MeV loses an a appreciable fraction of energy by generating continuous x-rays. The cross section for this process increases with the increase of energy E.

2.3 Gamma-rays

The interaction of γ-rays with the matter is remarkably different from that of the charged particles. The γ-rays are extremely penetrating and they can penetrate through a considerable thickness of the matter. They show an exponential absorption in the matter. Suppose a radiation of intensity I is incident upon an absorbing layer of thickness $\mathrm{d}x$, the amount of radiation absorbed $\mathrm{d}I$ is proportional to I as well as $\mathrm{d}x$, so that

$$\mathrm{d}I = -\mu I \, \mathrm{d}x \qquad \text{or} \qquad I = I_0 \, \mathrm{e}^{-\mu x}$$

where μ is known as the absorption coefficient and I_0 the initial intensity of radiation. At low energies (0.1 MeV to 25 MeV), there are three important processes through which the γ-photons are absorbed by the material: (i) Photoelectric effect, (ii) Pair production and (iii) Compton effect.

(i) Photoelectric effect

In this process, a γ-photon of proper frequency knocks out an electron from inside an atom of the absorbing material. The energy of emitted electron (photoelectron) is

$$\frac{1}{2} mv^2 = h\nu - W$$

where m and v are the mass and velocity of photoelectron, ν the frequency of incident γ-radiation and W the work function of the material. Thus, total energy of photon is taken; it is partly used as the work function of the material and rest is taken by the kinetic energy of photoelectron. (Here, after interaction, the photon does not exist any more.)

(ii) Pair production

In this process, the γ-radiation converts into a pair consisting of electron and positron. (Here, the energy is converted in to mass, following the law of energy-mass conversion.) This process can take place only when the photon energy exceeds $2m_0c^2$, where m_0 is the rest-mass of electron. The pair production cannot take place in a free space and generally takes place in the presence of a nuclear field. When the photon energy is larger than $2m_0c^2$, the balance energy is shared as the kinetic energy of the particles generated. (Here, also, after interaction, the photon does not exist any more.)

(iii) Compton effect

It is an elastic scattering process in which the incident photon gives a fraction of energy to an electron. When a photon of frequency ν incident upon a free electron at rest, the photon is scattered with less energy, which is a function of the angle θ which is the direction of deflection of the photon. The energy absorbed by the electrons is only a small fraction of the energy of incident γ-radiation. (Here, after interaction, the photon exists with less energy.)

At low energies of photon, the photoelectric effect is the main source of energy loss. With the increase of the energy of photon, the photoelectric effect becomes less effective. In the light-mass elements, Compton scattering becomes dominant when the photon energy is a few tens of KeV whereas for the heavier elements, it is not the case until the photon energy is nearly 1 MeV. The probability of the pair production becomes appreciable when the photon energy exceeds the threshold of $2m_0c^2 =$ 1.02 MeV.

3. Ionization chamber

The principle of ionization chamber is based on the fact that an energetic charged particle or γ-radiation ionizes the material of the medium through which it traverses. The apparatus consists of a cylindrical conducting chamber containing a central conducting electrode insulated from it (Figure 2). The gases used in it are generally the air or hydrogen at one atmospheric pressure, or at greater pressure for γ-radiation detection. One electrode of the chamber is connected to the high voltage supply so that its potential is V_0 and the other is connected through a resistance R.

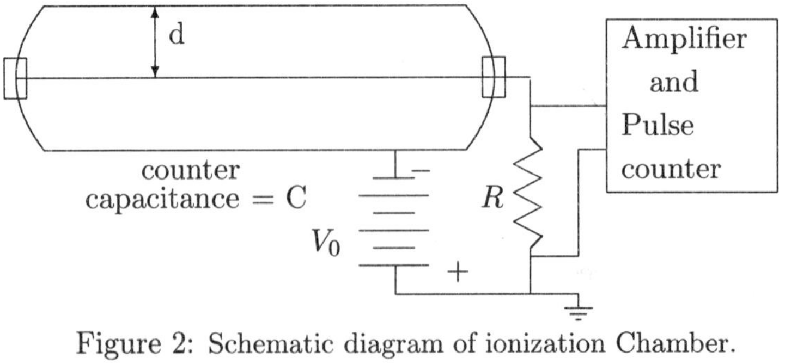

Figure 2: Schematic diagram of ionization Chamber.

When radiation enters into the chamber (through window), a certain number of electron-ion pairs are produced either directly (in case of charged particle) or by some secondary reaction (in case of neutral particle). The average number of pairs created is proportional to the energy deposited in the counter. Under the action of the electric field, the electrons are accelerated toward the anode and the ions towards the cathode, where they are collected, giving rise to a current signal that depends on the field intensity.

At the surface of anode, there will be high concentration of negative ions. Recombination in this region will be small as the rate of recombination is proportional to the product n^+n^-, where n^+ and n^- are the concentrations of positive and negative ions. Here, n^+ is very small. On the other side, at the cathodes, there will be high concentration of positive ions. Recombination in this region also will be small as n^- is very small there. Thus, the recombination will be maximum at the center of the space. A small current will flow through a resistor R connecting the electrodes externally. The size of this current pulse will depend upon the number of initially produced ions and would obviously be affected with the presence of the recombination process.

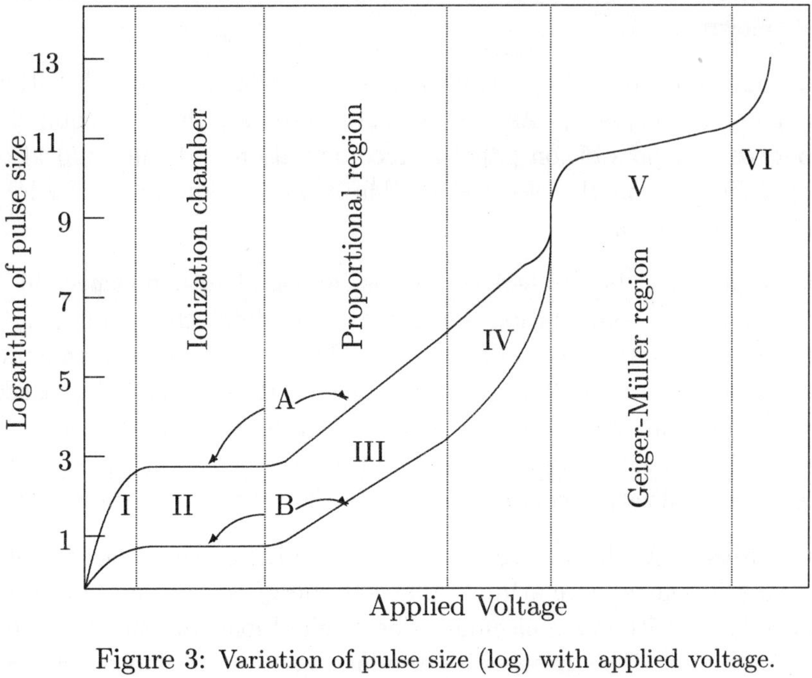

Figure 3: Variation of pulse size (log) with applied voltage.

With the increase of the voltage between the electrodes, the ions will move faster, the number of recombinations will be minimum and the pulse size will increase. The pulse size increases with the voltage until it reaches to a value which corresponds the state of no recombination of ions. This value of current is known as the saturation current and the chamber is said to reach a saturation. Further increase in the voltage will not change the pulse size. The range of voltage over which the pulse size remains constant depends on many factors, such as the nature and pressure of the gas, shape and size of the electrodes. The pulse size depends on the nature of the ionizing event.

The ionization chamber works in the region of constant pulse size. In this region the applied voltage is so high that recombination of ions is negligible.

With the further increase of the applied voltages, the electrons acquire sufficiently high energy and the secondary ionization initiates. Now, the ion current or pulse height increases above the saturation value. The size of the current pulse depends on the number of initial ion pairs produced by the ionizing particles and on the applied voltage. For a given ion pair, the pulse size is as shown in Figure 3 on a logarithmic scale as a function of the applied voltage.

An examination of this graph for variation of pulse size versus the applied voltage shows that the curve can be divided into six main region as the following.

In region I, the voltage applied is so low that the recombination of ion pairs takes place. As the voltage increases, there is competition between the loss of ion pairs by recombination and the removal of ions by collection on the electrodes. The region is known as the ion-recombination region.

In region II, the applied voltage is increased to sufficiently high value so that the recombination is negligible and all the charged particles produced move to the electrodes rapidly. Thus, the pulse-size remains independent of the potential applied as long as the voltage is increased up to the point at which the secondary electrons are produced by the collisions of primary electrons with the atoms of the gas. This region is known as the ionization chamber region or the saturation region.

In region III, the voltage is sufficiently high so that electron approaching the central wire attains sufficient energy to produce new ions, due to collision with gas molecules. The original ions and the new ones combine together. This process is cumulative and an avalanche of electrons is thus produced. The number of the ion-pairs formed by collision as each electron travels towards the central wire is known as the multiplication factor M. The voltage region in which M remains constant is known as the proportional region. In this region, the pulse height remains proportional to the amount of energy lost in the chamber by the primary ionizing particle.

In region IV, the voltage is increased further and the pulse height continues to increase. The height is still related, but no longer proportional to the initial ionizing intensity. This voltage region where M depends some what on the pulse size is known as the region of limited proportionality.

In region V, the pulse height is very large and also substantially independent of the initial number of ions. This is known as the Geiger region and the device which operates in this region is known as the Geiger-Müller counter. In the Geiger region, the avalanche associated with the drift of a single electron to one point on the anode initiates avalanches over the entire anode length.

In region VI, the increase in voltage gives a continuous breakdown with or without radiation. This region is generally avoided, in order to prevent the damage of the detector. That is, this region is not used for the purpose of any detector.

Ionization chambers are used widely for the study of α-particles, β-particles, protons, deuterons and the nuclei of light elements. They can also be used for the measurement of γ-radiation and x-rays. In this case, the ionization chambers with walls of elements like Platinum or Bismuth with a high atomic number are used to prevent the release of electrons from the walls of the chamber.

Exercise 1: α-particles, each having energy 5 MeV, pass through an ionization chamber at the rate of 10 per second. Assuming that all the energy is used for the production of ion pairs, calculate the current produced. Given, 35 eV is required energy for the production of one ion pair; charge of electron $e = 1.6 \times 10^{-19}$ C.

Solution: Energy of one α-particle $= 5 \times 10^6$ eV

Energy required to produce one ion-pair $= 35$ eV

Number of ion-pairs produced by one α-particle is

$$= \frac{5 \times 10^6}{35} = 1.429 \times 10^5$$

As 10 particles enter the chamber in one second. Total number of ion-pairs produced $= 1.429 \times 10^5 \times 10 = 1.429 \times 10^6$

Charge on each ion $= 1.6 \times 10^{-19}$ C.

Hence, the current $= 1.429 \times 10^6 \times 1.6 \times 10^{-19} = 2.286 \times 10^{-13}$ A

4. Proportional counter

Proportional counters are of great use in the detection of low energy x-rays, electrons and also, with certain operating conditions, of neutrons. When a particle of low energy passes through an ionization chamber, the pulse produced is too small to detect. Potential difference is increased so that it works in the proportional region, in which the size of the output signal is proportional to the number of ions formed by the primary ionization process.

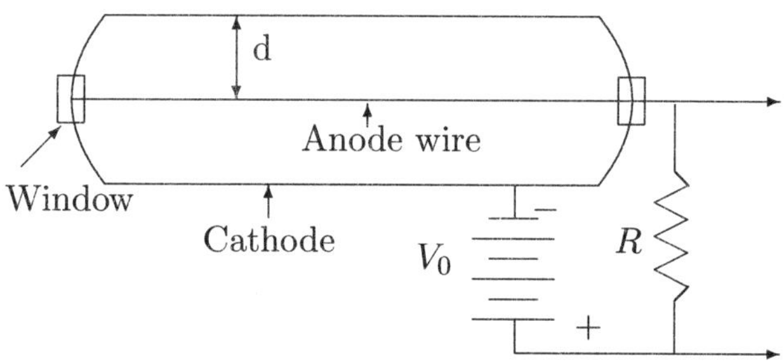

Figure 4: Schematic diagram of proportional counter.

A proportional counter consists of a cylindrical gas filled tube with a very thin central wire, which is insulated from the tube. The central wire (diameter of about 0.2 mm) is given positive potential with respect to the metallic cylindrical tube. It serves as a collecting electrode and is connected to a pulse amplifier.

The electrons ejected from the ionization of gas molecule drift towards the positive electrode. When they move in a region of high electric field E, they will produce additional ionization by collisions with molecules of the gas. The electric field strength is

$$E = \frac{V_0}{x \ln(b/a)}$$

where x is the radial distance from the axis, b the radius of the cylinder and a the radius of the central wire and V_0 the positive voltage applied to central wire. The field strength decreases with the increase of x.

For $V_0 = 1000$ volts, $b = 1.0$ cm, $a = 0.1$ mm, the field strength at the wire is

$$E = \frac{1000}{0.1 \times 10^{-3} \ln(10^{-2}/0.1 \times 10^{-3})} = 2.17 \times 10^6 \text{ V/m}$$

Application of high field will cause the multiplication of ionization process, resulting avalanche and high signal is generated. The choice of gas filled in the proportional counter is very important and it depends on the following factors: low working voltage, high gain, good proportionality and high capability. These requirements are generally obtained by filling a mixture of gases instead of a single gas. Nobel gases (e.g., Ar) are generally chosen, as they require the lowest electric field intensities for the avalanche formation.

The gain of the proportional counter can be calculated in a simple manner as the following. Suppose λ is the mean free path of electron for a secondary ionizing collision, and $1/\lambda$ is the probability of ionization per unit path length. When there are N number of electrons, then in the path length $\mathrm{d}x$, the number of new electrons created is

$$\mathrm{d}N = N\alpha\,\mathrm{d}x \tag{7.1}$$

Here, α is a constant. Solution of equation (7.1) is

$$N = N_0 \exp(\alpha x)$$

where N_0 is the initial number of electrons. Therefore, the multiplication factor is

$$M = N/N_0$$

It depends on the anode diameter, the radius of the cathode, potential difference and the nature and pressure of the gas.

Exercise 2: A proportional counter is required to operate with a maximum radial field of 5×10^6 V/m. Calculate the applied voltage required when the radii of central wire and cylindrical tube are 0.001 cm and 1 cm, respectively.

Solution: The radial field strength E is expressed as

$$E = \frac{V_0}{x\,\ln(b/a)}$$

where V_0 is the applied voltage, b the radius of the cylindrical tube, a the radius of the central wire and x is the radial distance from the axis of the tube. The field is maximum at $x = a$. On substituting the given values of various parameters, we get

$$5 \times 10^6 = \frac{V_0}{10^{-5}\,\ln(10^{-2}/10^{-5})}$$

On solving, we get $V_0 = 345.4$ V.

5. Solid state detector

K.G. Mckay was the first who showed in 1949 that a *p*-*n* junction may be used as a detector. He allowed a collimated beam of α-particles to fall on a *n*-*p* germanium junction and showed that nearly whole of the energy of α-particles absorbed in it, was utilized for production of electron-hole pairs. The important aspect has been that the average energy required to produce one electron-hole pair was about 3 eV, which is much small in comparison to 35 eV energy required for the formation of one ion pair in the gas. A *p*-*n* detector is shown in Figure 5.

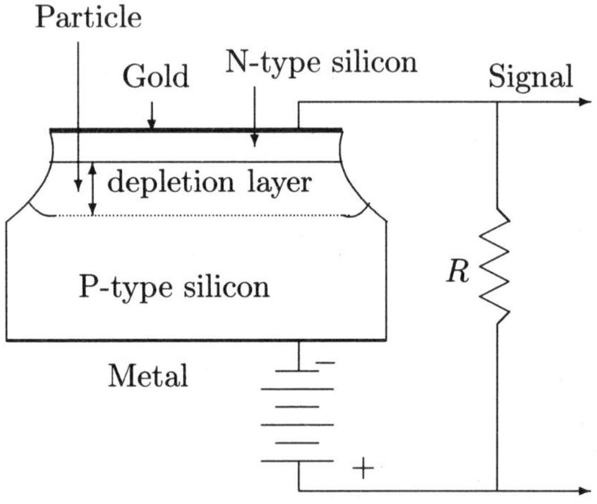

Figure 5: A *p*-*n* junction detector.

It consists of a *p*-*n* junction between *p*-type and *n*-type silicon. The *n*-type silicon layer is made in contact with a thin film of gold developed through the evaporation technique. The reverse biased diode helps in minimizing the current flowing in the detector when no radiation is striking the detector. A positive (reverse) bias applied to the gold film pushes all the positive charge carriers away from the junction and produces a depletion layer, shown in Figure 5. The depletion layer contains almost no positive or negative charge carriers. When an energetic charged particle moves through the depletion layer, its interaction with the electrons in the crystal, produces electron-hole pairs. One electron-hole pair is produced for every 3.5 eV (in Si) energy, lost by the charged particle. The electrons and holes are swept away by the applied electric field and are registered as a voltage pulse over the register R. The number of charge carrier pairs produced in a semiconductor material is approximately 10 times as large as the number of ion pairs produced in a gas ion chamber. The voltage pulse will therefore be about 10 times large. Thus, this detector has better resolution of energy than the radiation detectors.

In the solid state detectors for charged particles, silicon has been generally used because of its low intrinsic conductivity. This shows that the detector can be operated at room temperature without excessive leakage of current. For the detection of γ-rays, germanium detectors are much better than those of silicon, because of large density of germanium.

6. Geiger-Müller counter

Geiger and Müller in 1928 designed a counter, known as the Geiger-Müller counter (or G-M counter), for detection and counting of radioactive radiation. It consists of a central electrode A, made of tungsten wire of about 0.03 - 0.2 mm diameter, stretched along the axis of a metal tube B (Figure 6). The metal tube is well insulated from the central electrode with the help of ebonite plugs E and F. A high tension battery (about 1000 to 3000 V) is applied between A and B though a high resistance R (more than 100 megaohms). The negative is connected to B and positive to A. The tube is filled with a gas mixture of about 90 % argon and 10 % alcohol at a pressure of about 10 cm of Hg. The DC voltage is kept slightly less than that for which there is a discharge between the electrodes.

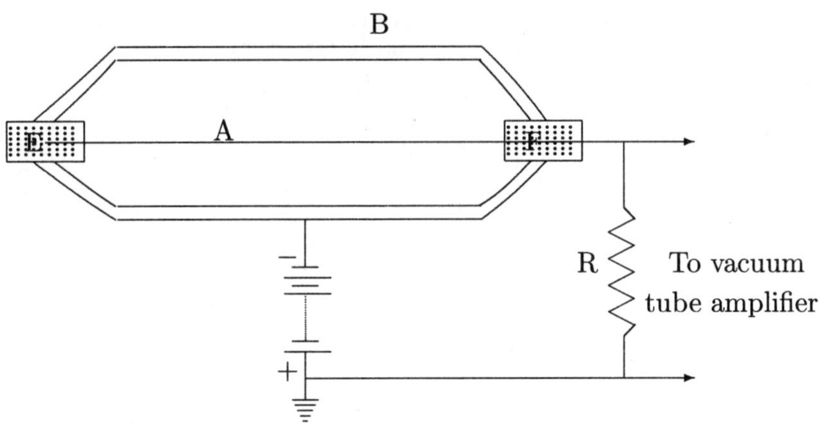

Figure 6: Schematic diagram of Geiger-Müller counter.

When an ionizing particle (*e.g.*, an α-particle) enters the counter, ionization takes place and some ions are produced. When the potential difference is high enough, the ions are multiplied by further collisions. An avalanche of electrons moves towards the central wire A and this is equivalent to a small current impulse which flows through the load resistance R. The critical potential is lowered momentarily, causing a sudden discharge through the resistance R. The potential difference so developed across R is amplified by vacuum tube circuit and is made to operate a mechanical counter. This way a single particle can be registered. A sudden pulse of discharge sweeps away the ions from the chamber and the counter is ready to register the arrival of the next particle.

The voltage characteristic of G-M counter are shown in Figure 7. This is a plot of the counting rate versus the applied voltage with a radioactive source placed near the counter. It can be seen that there is a threshold voltage below which the tube does not work. It can be several hundred volts. When the applied potential increases, the counting begins and rises rapidly to a flat portion of the curve, called the plateau. This is the region where the G-M counter works and the counting rate is more or less does not depend on the small changes in the potential difference across the tube. With the increase of potential difference, beyond the plateau, a continuous discharge takes place in the tube and the count rate increases very rapidly. It does not require any ionizing event for this to take place and the tube must not be used in this region.

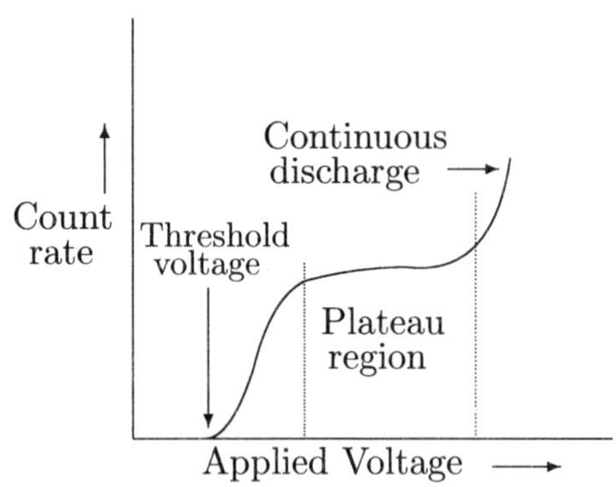

Figure 7: Voltage characteristics curve of G-M counter.

The efficiency of the counter is defined as the ratio of the observed counts per seconds to the number of ionizing particles entering the counter per second. The counting efficiency is defined as the ability of its counting if at least one ion-pair is produced in it. The efficiency ϵ is expressed as

$$\epsilon = (1 - e^{-slp})$$

where s is the specific ionization at one atmosphere, p the pressure in the units of atmosphere and l the path length of the ionization particle in the counter.

Disadvantages of G-M counter are: (i) it is insensitive for a period of 200 to 400 μs following each pulse. It prevents its use at very high counting rates. (ii) it cannot provide information about the particle or radiation causing the pulse.

Exercise 3: An organic quenched G-M tube operates at 1000 volts. Its anode is a wire of radius 0.1 mm and the cathode is a tube of radius 2 cm. The tube has a guaranteed life time of 10^9 counts. What is the maximum radial field and how long will the counter last if it is used on the average for 30 hours per week at 2000 counts per minute? Consider 50 weeks in one year.

Solution: The radial field strength E is expressed as

$$E = \frac{V_0}{x \ln(b/a)}$$

where V_0 is the applied voltage, b the radius of the cylindrical tube, a the radius of the central wire and x is the radial distance from the axis of the tube. The field is maximum at $x = a$. On substituting the given values of various parameters, we get

$$E = \frac{1000}{1 \times 10^{-4} \ln(2 \times 10^{-2}/1 \times 10^{-4})} = 1.89 \times 10^6 \text{ V/m}$$

If the life time of the tube is N years, it is equal to 50 N weeks.

The tube is used for 50 $N \times 30 = 1500\ N$ hr $= 9 \times 10^4\ N$ min

No. of count in $9 \times 10^4\ N$ min $= 9 \times 10^4\ N \times 2000 = 1.8 \times 10^8\ N$ counts

These counts must be equal to the guaranteed life time of 10^9 counts. Thus, we have

$$1.8 \times 10^8 N = 10^9 \qquad \text{or} \qquad N = 5.556 \text{ years}$$

7. Nuclear emulsions

A photographic plate plays important role in the investigation of charged particles as well as of the radiation. A photographic plate carries a coating of a material, called the emulsion. A heavy charged particle traversing through a photographic emulsion produces a latent image of its track. This track is revealed when the plate is developed. Since the sensitivity of ordinary optical photographic emulsions is low, they are not suitable for the quantitative studies of nuclear particle/radiation. Moreover, the tracks due to charged particles have non-clear range as the developed crystal grains are large and widely spaced. In order to make the emulsion more suitable for the study of nuclear radiation, such as α-particles, protons, mesons and even electrons, its composition needs to be changed. The nuclear emulsions differ from the optical ones in the sense that they have considerably higher silver halide content and smaller grain size. In nuclear emulsions, the thickness is larger than that of the optical emulsion. The smaller grain size, the more sensitive is the emulsion to the ionizing particles. Thus, commercially available, various emulsions differing in grain size, can be used in order to distinguish between different particles.

7.1 Advantages

The use of emulsions is accompanied with some advantages. Some of them are as the following.

1. The emulsions are relatively light and cheap. Because of their light mass, they can be sent in the balloons, spaceships etc. for performing high altitude experiments for cosmic rays. After recording of a cosmic ray event, the plate is brought down to the earth's surface and developed in a laboratory. Emulsions have been used widely in the cosmic ray experiments.

2. High density of an emulsion gives a stopping which is power about a thousand times that of the standard air. Unstable high energy particles are brought to rest in the emulsions thus their decay schemes are studied.

3. The emulsions are continuously sensitive and therefore they are always available for recording an event. On the other side, for example, the cloud chamber is sensitive only for a fraction of second and remains ineffective for several seconds before becoming sensitive again.

7.2 Limitations

Though the use of emulsions has advantages, it however has some limitations. Some of them are as the following.

1. The main drawback of nuclear emulsions is that their sensitivity and thickness are affected by the temperature, humidity, age of emulsion before development and the conditions under which they are developed.

2. The scanning of the plates and the analysis of the tracks obtained are very laborious job, when they are done manually.

3. In case, of nuclear emulsions, it is very difficult to determine the sign of the charge and the momentum of a particle from observations of the curvature of the path in a strong magnetic field. The actual path-length in the emulsion is small as compared to that in a cloud chamber.

8. Scintillation detector

Some materials have scintillation property and because of that they emit light when a nuclear particle or radiation incident on any of them. This small flash of light can be amplified with the help of a photomultiplier tube and is finally converted into an electrical pulse, which can be analyzed and counted electronically to provide information about the incident radiation. The most commonly used scintillators include the inorganic alkali halides like sodium iodide and organic based liquids and plastics. Inorganic crystals give better light output with good linearity, so that the light yield should be proportional to the deposited energy over a wide range. However, the response of these materials is poor.

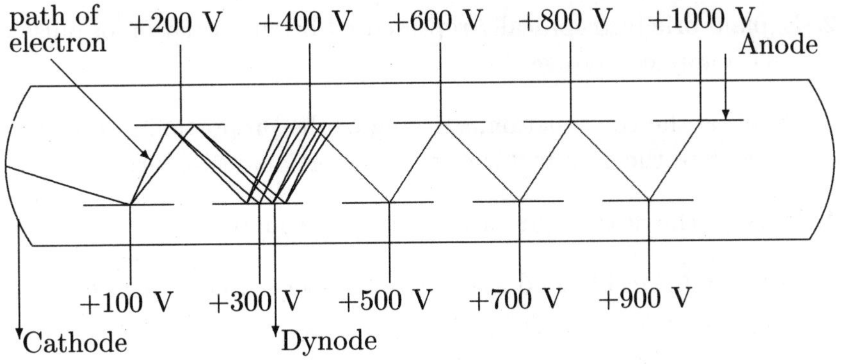

Figure 8: Schematic diagram of photomultiplier tube.

Scintillation detector consists of a scintillating material, which is optically coupled to a photomultiplier either directly or via light guide.

The light coming out from scintillator due to the impact of radiation is transmitted to the photomultiplier where it is converted into a weak current of photoelectrons, which are then further amplified by an electron multiplier system. The resulting current signal is then analyzed by an electronic system.

The next stage of scintillation detector system is the photomultiplier tube, which converts the light into a measurable electric current. The basic components of the photomultiplier tube are schematically shown in Figure 8. It consists of cathode made of photo sensitive material, followed by an electron collection system, an electron multiplier section, and finally an anode from where the final signal is to be taken. A block diagram about all steps is shown in Figure 9. All parts are usually housed in an evacuated glass tube. The photo-cathode converts incident light into a current of electrons by the photoelectric effect.

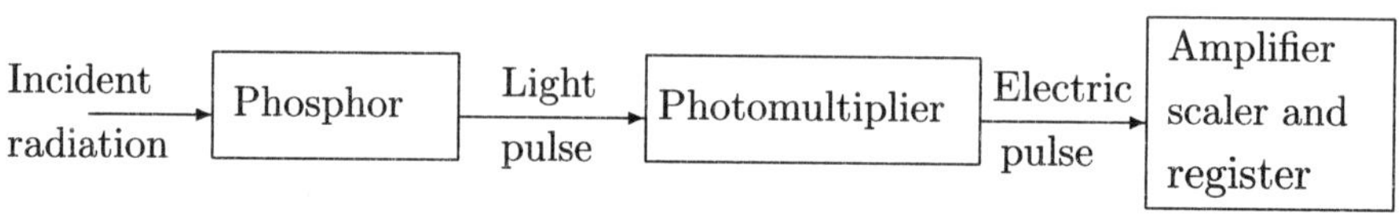

Figure 9: Block diagram of working of Scintillation detector.

9. Problems and questions

1. Describe the construction, working and theory of the Ionization detector.
2. Explain briefly graphically represented by the variation of pulse size with applied voltage.
3. Describe the construction, working of the proportional counter and calculate the gain factor of the proportional counter.
4. What is the advantages of proportional counter?
5. What is mean by solid state detector and explain in brief diffused junction detector.
6. Describe the construction, working and characteristics curve of the G-M counter.
7. Explain the construction and working of the photomultiplier tube based scintillation detectors.
8. Write short note on the following:

   (i) G-M counter

   (ii) Ionization chamber

(iii) Proportional counter

(iv) Solid state detector

(v) Nuclear emulsions

(vi) Scintillation detector

VIII. Accelerators

For the study of nuclear reactions, scientists often require incident particles moving with various energies. Since the particles involved in the nuclear reactions carry electric charge, their energies can be increased with the help of electric field or with the help of varying magnetic field. An apparatus designed for this purpose is known as the particle accelerator. Thus, a particle accelerator is a device which is used to increase kinetic energy of electrically charged particles. The particles generally used for acceleration are electrons and nuclei of light elements, such as protons, deuterons and alpha particles. However, heavier nuclei, such as carbon, beryllium, nitrogen, oxygen and neon also have been accelerated. Now, heavy ions such as silicon, nickel, iodine and lead are commonly used in the advanced level accelerators. Obviously, the particle accelerators play important role in carrying out research in the field of elementary particles, and for the study of the structure of a nucleus, nuclear forces, and strong and weak interactions. For this reason, they are often known as the nuclear accelerators.

When the electric field is applied on the charged particles obtained from a source, the particles are accelerated along the direction of the field. Positive charges are accelerated along the direction of the field, whereas the negative charges in the opposite direction. On the other side, a magnetic field exerts a force at right angle to the plane containing the the applied magnetic field and the direction of motion of the charged particle. Hence, a magnetic field does not do any work on a moving charged particle and consequently, the magnetic field cannot change its energy. A uniform magnetic field can produce circular orbits of charged particles. But, an increasing magnetic field can accelerate the charged particles.

1. Electrostatic accelerators

In an electrostatic accelerator, a high voltage is developed between two terminals and the charged particles to be accelerated are allowed to travel along the path between the terminals. Suppose V is the potential difference developed between these two terminals separated by a distance d, then the electric field E between them is $E = V/d$. Now, a particle carrying a charge Ze experiences a force $F = ZeE$ along the direction of the electric field. Here, e denotes the charge of an electron. Positive charges are accelerated along the direction of the field, whereas the negative charges in the opposite direction. Hence, the charged particle is accelerated by the potential difference between the two terminals. After transversing between the terminals, the work done $W = Fd = ZeV$. Hence, the energy gained by a particle during its motion between the terminals is ZeV joules or ZV eV. There are some electrostatic accelerators which produce a potential difference in the range of million volts. One of the electrostatic accelerators is the van de Graaff generator.

1.1 Van de Graaff generator

In 1929, Van de Graaff designed a particle accelerator which can produce a potential difference up to 5 million volts. Its design is based on the principle that when a conductor having charge is brought into the internal contact with another hollow conductor, all of its charge can be transfered to the hollow conductor. Theoretically, the potential of the hollow conductor can be increased to any desired value by successive addition of charge to it. But, in practice, its maximum potential is limited to that at which the rate of leakage of charge from it is equal to the rate at which the charge is delivered to it.

A simple schematic diagram of Van de Graaff generator used to accelerate positive ions is shown in Figure 1. It consists of a belt of silk or a similar flexible dielectric material, running over a pair of metal pulleys, P and Q. The upper pulley P is surrounded by a hollow metal sphere S. Two electrodes, A and B, in the form of comb-shaped rows of sharp metal points, are placed respectively, near the bottom of the lower pulley, and top of the upper pulley. Comb B is connected to the sphere S whereas the comb A to a DC supply.

In case of positive potential, a high DC potential (with respect to earth) is applied to comb A. The motion of the belt is shown by arrows. As the belt passes in front of the comb A, it receives positive charge which escapes from the sharp metal points of the comb A. As the belt is made up of insulator material, the positive charges remain stick on the belt which is moving in the upper direction. When the belt reaches in front of the comb B, it attracts negative charge from the atmosphere, and consequently a positive charge is given to the comb B. This charge is transfered to the metal sphere.

This process for transfer of charge goes on continuously and the potential of the sphere goes on increasing until the rate of loss of charge by leakage becomes equal to the rate at which the charge is transfered to the sphere. The leakage of charge can be minimized and the maximum attainable voltage can be increased by enclosing the apparatus in a gas-tight steel chamber, operating at the pressures up to about 15 atmospheres. The gas in the chamber may be air, nitrogen, methane, or freon.

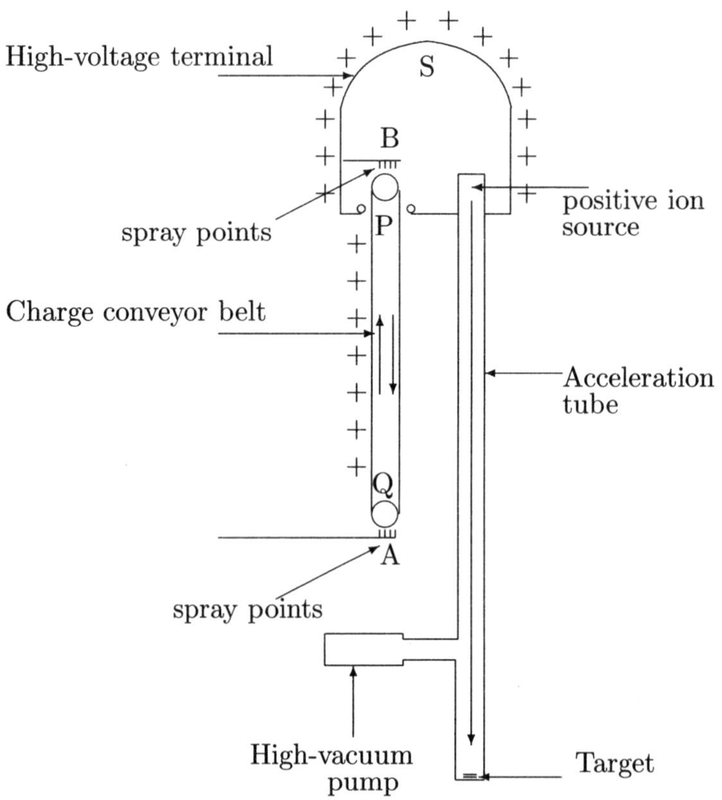

Figure 1: Schematic diagram of a Van de Graaff generator used to accelerate positive ions.

A gas discharge tube, working as a source of positive ions, is inside the shell. The ions are accelerated in the downward direction and strike the target located at the low voltage end of the tube. The accelerating tubes are generally made up of sections of glass, porcelain, or other insulating material joined end to end by the vacuum-tight seals, and must be long enough to avoid the possibility of spark during passing of ions from one end to the other, when the potential difference is applied.

2. Cyclotron

Lawrence and Livingston developed a heavy ion accelerator which is based on the principle that the particles moving in a magnetic field can be accelerated by using one pair of electrodes only where the particles are allowed to cross the same accelerating gap repeatedly. This instrument is known as the 'magnetic resonance accelerator' or 'cyclotron'.

Schematic diagram of a cyclotron is shown in Figure 2. It consists of two flat, semicircular metal boxes, called 'dees' because of their shape. These hollow chambers have their diametric edges parallel and slightly separated from each other. A source of ion is kept near the mid-point of the gap between the dees. The dees are connected to the terminals of an oscillator having frequency of several million cycles per second. The dees behave as electrodes where alternating potential is applied. In this way the potential between the dees is made to alternate rapidly, some millions of times per second, and the electric field in the gap is directed first toward one of the dees and then towards the other dee. The space within each dee is, however, a region of zero electric field because of the electrical shielding effect of the dees. The dees are enclosed within a larger metal box containing gas at low pressure, and the whole apparatus is placed between the poles of a strong electromagnet which provides a magnetic field perpendicular to the plane of the dees.

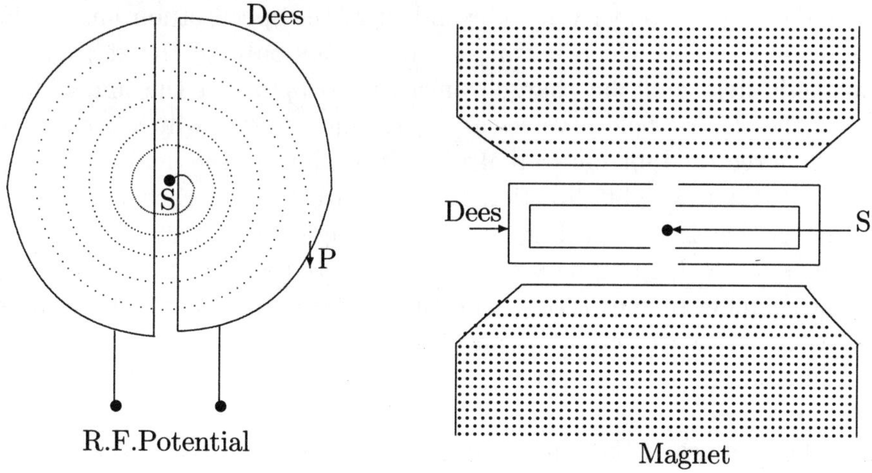

Figure 2: Cyclotron (Magnetic resonance accelerator).

When an ion with a positive charge q is emitted from the source, it is accelerated by the electric field between the dees toward the dee which is negative at that instant. Since there is a uniform magnetic field B

acting at right angles to the plane of the dees, the ion moves in a circular path of radius r_1, expressed as

$$r_1 = \frac{mv_1}{Bq}$$

where m is the mass of the ion, B is the magnetic field strength and v_1 is the speed of the ion.

When the ion is inside the dee, its speed remains constant. After describing a semicircle through the dee, the ion reaches the gap, where it is again subject to the action of the applied potential difference. Now, when the time needed for the ion to complete the semicircle is half of the time-period of the applied electric field, the direction of the field is toward the second dee. Hence, the ion is again accelerated as it crosses the gap between the dees and enters the second dee with an increased speed v_2. It then moves within the second dee in a semicircle of larger radius r_2 expressed as

$$r_2 = \frac{mv_2}{Bq}$$

The angular velocity of the ion is

$$\omega = \frac{v}{r} = \frac{qB}{m} \tag{8.1}$$

where v is the velocity of ion moving in a semicircle of radius r. It shows that the angular velocity is independent of the speed of ion and of the radius of circle in which it travels. It depends only on the strength of magnetic field and the charge-to-mass ratio (q/m) of the ion. Thus, when the electric field reverses at the regular interval which is equal to the time required for ion to make a half revolution, the field in the gap is always in the right direction to accelerate the ion each time. The radius of the circular path goes on increasing continuously. Eventually, the ions reach the periphery of the dees, where they can be brought out of the chamber by means of a deflecting plate charged to a high negative potential. The attractive force acting on the positive ions draws them out of their spiral paths. These ions can be used to bombard a properly placed target.

The energies of the ions produced by a cyclotron can be calculated from the equation describing the motion of an ion in a magnetic field:

$$Bqv = \frac{mv^2}{R} \tag{8.2}$$

where R is the radius at which the particle leaves the machine, and the magnetic field B and charge q are expressed in the electromagnetic units.

After an ion has been accelerated, its energy is

$$qV = \frac{1}{2}\, mv^2 \tag{8.3}$$

where V is the equivalent potential difference through which the ion is accelerated. Using the value of v from equation (8.2) in (8.3), we get

$$V = \frac{1}{2}\, B^2R^2\, \frac{q}{m} \tag{8.4}$$

Hence, when the radius is doubled, the equivalent voltage through which the ions are accelerated is increased by a factor of four.

For resonance the time taken for an ion, traveling with an angular velocity ω, to describe a semicircle within one of the dees must be equal to half of the time-period T of the applied high frequency voltage. Thus, we have

$$\frac{\pi}{\omega} = \frac{T}{2} \qquad \text{or} \qquad T = \frac{2\pi m}{qB}$$

and the frequency f is

$$f = \frac{1}{T} = \frac{\omega}{2\pi} = \frac{qB}{2\pi m}$$

For a given charge, for higher frequency, the magnetic field be stronger, and therefore from equation (8.4), the equivalent voltage through which the ions will be accelerated at any given radius will be larger. The energies to which particles can be accelerated in a cyclotron are limited by the relativistic increase of mass with velocity. When the velocity of a particle is comparable to the speed of light, its mass varies as

$$m = \frac{m_0}{\sqrt{1 - (v^2/c^2)}}$$

where m_0 is the rest-mass of the particle. Then from equation (8.1), the angular velocity of ions becomes

$$\omega = \frac{qB}{m_0}\, \sqrt{1 - (v^2/c^2)} \tag{8.5}$$

Obviously, in the relativistic limit, the angular velocity decreases with the increase of velocity. Further, the frequency of ions rotation decreases. That is, the ions take longer time to describe their semicircular paths as compared to the fixed period of the oscillating electric field and they reach late at the gap between the dees. Consequently, they lag behind in phase the voltage applied to the dees until finally they are no longer accelerated. The ion velocities at which the relativistic mass increase leads to this effects represents a limiting size and energy for the cyclotron.

Exercise 1: A cyclotron in which the flux density is 1.4 Wb/m^2 is used to accelerate protons. How rapidly should the electric field between the dees be reversed? Given, mass of proton = 1.67 $\times 10^{-27}$ kg and charge of proton = 1.6 $\times 10^{-19}$ C.

Solution: We have $B = 1.4$ Wb/m^2, $m = 1.67 \times 10^{-27}$ kg and $q = 1.6 \times 10^{-19}$ C. Thus, we have

$$t = \frac{\pi m}{Bq} = \frac{3.14 \times 1.67 \times 10^{-27}}{1.4 \times 1.6 \times 10^{-19}} = 2.341 \times 10^{-8} \text{ s}$$

Exercise 2: Deuterons in a cyclotron describe a circle of radius 1.32 m just before emerging from the dees. The frequency of the applied voltage is 10 MHz. Calculate the flux density of the magnetic field and the velocity of deuterons emerging out of the cyclotron. Given, mass of deuteron = 3.32 $\times 10^{-27}$ kg.

Solution: We are given, $m = 3.32 \times 10^{-27}$ kg, $f = 10 \times 10^6$ Hz, $q = 1.6 \times 10^{-19}$ C. Now, we have

$$B = \frac{2\pi m f}{q} = \frac{2 \times 3.14 \times 3.32 \times 10^{-27} \times 10^7}{1.6 \times 10^{-19}} = 1.303 \text{ Wb/m}^2$$

The velocity v is

$$v = \frac{Bqr}{m} = \frac{1.303 \times 1.6 \times 10^{-19} \times 0.32}{3.32 \times 10^{-27}} = 2.009 \times 10^7 \text{ m/s}$$

3. Synchro-cyclotron

For the relativistic case, the frequency f can be expressed as

$$f = \frac{qB}{2\pi m_0} \sqrt{1 - (v^2/c^2)} \qquad (8.6)$$

Obviously, with the increase of velocity v, the frequency n decreases and the resonance between the applied field and the motion of charged particle is lost. The loss of resonance can however be balanced in two ways:

1. Equation (8.6) shows that to maintain the resonance, the frequency of the applied voltage should be gradually decreased, so that it always matched with the frequency of circulation. Accelerators employing this principle are called frequency modulated cyclotrons or synchro-cyclotron.

2. Another possibility is that the magnetic field is decreased in such a manner so that $B\sqrt{1-(v^2/c^2)}$ remains constant. Accelerators working on this principle are also called synchro-cyclotron.

It was thought for some years that from the experimental point of view, it would not be possible to design a machine which could satisfy either of the above conditions. But, it was shown theoretically later on that these methods could be applied successfully and that the possibility depends on a property called the phase stability possessed by certain orbits in a cyclotron.

It is possible to construct a variable frequency cyclotron where phase stability and be obtained. It can be understood in a qualitative manner by considering motion of charged particles. Consider an ion moving in a circular path in a uniform magnetic field in such a manner that on each revolution it crosses the gap between the accelerating electrodes to which is applied an oscillating electric field whose frequency is identical with the frequency of rotation of the ion. Under such conditions there are certain 'stationary' or stable orbits in which the ion can move. Let us first consider the ions which cross the gap at instants of time when the oscillating electric field is just passing through the value zero. The ions are then said to have zero phase, and this condition is obtained at the points 0, 2π, 4π and so on in Figure (3); these ions will neither gain nor lose energy and will continue to revolve at constant frequency and in the same orbit.

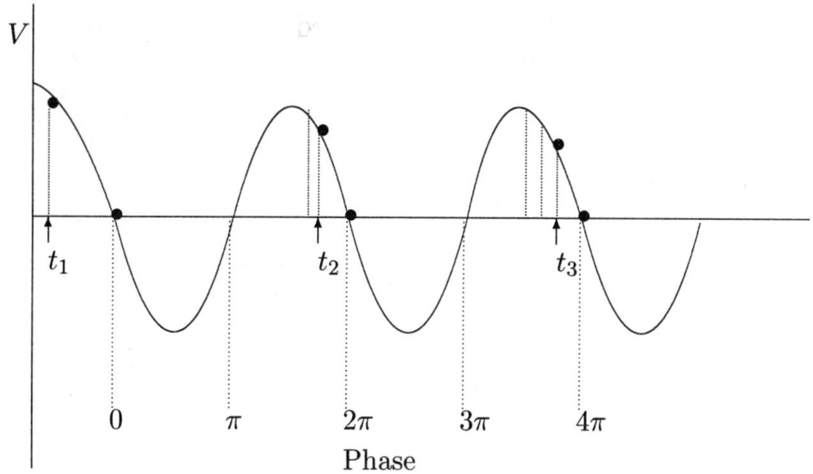

Figure 3: Phase stability in a cyclic magnetic accelerator.

An ion which crosses the gap at an earlier instant such as t_1 has a positive phase; it will gain speed and energy since the voltage is positive (*i.e.*, in the acceleration phase) and its rotation frequency will decrease in accordance with equation (8.5). The ion will take a slightly longer time to return to the gap, as shown by points t_2 and t_3 in the subsequent acceleration phases, where the voltage across the gap is smaller. Finally, the particle will cross the gap at zero phase, but with the excess energy it has received, and will continue the phase shift into the decelerating part of the cycle when the voltage is negative (*i.e.*, in the deceleration phase). The particle then loses energy and speed, its frequency increases, and it is returned to the zero phase. There has been a phase oscillation about the equilibrium phase $\theta = 0$, and the ion has oscillated about a 'stationary' or equilibrium orbit.

By increasing the magnetic field or reducing the frequency of the dee-voltage, or both, the orbits of the ions could be made to expand the energy to increase. When the variation is made sufficiently slowly, the phase stability can be preserved during the acceleration, and the limits on the ordinary cyclotron can be passed. Particles can be accelerated by keeping the electric field frequency at a constant value and varying the magnetic field; in this case the machine is known as a synchrotron. Protons, deuterons, and α-particles can be accelerated to high energies by keeping the magnetic field constant and varying the frequency of the electric field; in this case the machine is known as a frequency modulated (FM) cyclotron or synchrotron.

Exercise 3: Calculate the accelerator voltage required for accelerating a proton to a velocity of 95% the speed of light. Given, rest mass of proton $m_0 = 1.672 \times 10^{-27}$ kg.

Solution: Mass of proton moving with velocity 95% the speed of light is

$$m = \frac{m_0}{\sqrt{1-(v^2/c^2)}} = \frac{1.672 \times 10^{-27}}{\sqrt{1-(95/100)^2}} = 5.355 \times 10^{-27} \text{ kg}$$

Thus, the change in mass $\Delta m = (5.355 - 1.672) \times 10^{-27}$ kg

Therefore, the kinetic energy

$$E = \Delta m\, c^2 = (5.355 - 1.672) \times 10^{-27} \times 9 \times 10^{16}$$

$$= 3.31 \times 10^{-10} \text{ J} = 2.05 \times 10^9 \text{ eV}$$

Hence, the accelerating voltage $=$ 2.05 BeV.

Exercise 4: In a synchro-cyclotron, the flux density decreases from 1.5 Wb/m^2 at the center of the magnet to 1.43 Wb/m^2 at the limiting radius of 2.06 m. Calculate the range of frequency modulation required for deuteron acceleration and the maximum kinetic energy of the deuterons. Given, the rest mass of deuteron is 3.34 $\times 10^{-27}$ kg; electron charge is 1.62 $\times 10^{-19}$ C.

Solution: We have the relation

$$Bqv = \frac{mv^2}{r} \quad \text{where} \quad m = \frac{m_0}{\sqrt{1 - v^2/c^2}} \quad \text{so that} \quad \frac{Bqr}{m_0} = \frac{v}{\sqrt{1 - v^2/c^2}}$$

Using $B = 1.43$ Wb/m^2, $m_0 = 3.34 \times 10^{-27}$ kg, $r = 2.06$ m, $c = 3 \times 10^8$ m/s, $q = 1.62 \times 10^{-19}$ C, we get

$$\frac{1.43 \times 1.62 \times 10^{-19} \times 2.06}{3.34 \times 10^{-27}} = \frac{v}{\sqrt{1 - v^2/9 \times 10^{16}}}$$

On solving, we get $v = 1.36 \times 10^8$ m/s. The frequency of revolution f is

$$f = \frac{v}{2\pi r} = \frac{1.36 \times 10^8}{2 \times 3.14 \times 2.06} = 10.51 \times 10^6 \text{ Hz} = 10.51 \text{ MHz}$$

The initial frequency of the oscillator must be equal to the simple cyclotron frequency

$$f = \frac{Bq}{2\pi m_0} = \frac{1.5 \times 1.62 \times 10^{-19}}{2 \times 3.14 \times 3.34 \times 10^{-27}} = 11.59 \times 10^6 \text{ Hz} = 11.59 \text{ MHz}$$

Thus, the range of frequency modulation is from 11.59 MHz to 10.51 MHz.

The kinetic energy is

$$E = m_0 c^2 \Big[\frac{1}{\sqrt{1 - v^2/c^2}} - 1\Big]$$

$$= 3.34 \times 10^{-27} \times 9 \times 10^{16} \Big[\frac{1}{\sqrt{1 - (1.277/3)^2}} - 1\Big]$$

$$= 3.16 \times 10^{-11} \text{ J} = 197.5 \times 10^6 \text{ eV} = 197.5 \text{ MeV}$$

4. Betatron and electron synchrotron

The Van de Graaff generator can be used to accelerate electrons, but the energy is limited to a few million eV. On the other side, an ordinary cyclotron cannot be used for electrons because of the large relativistic increase of mass at low energies. For example, an electron with an energy of 1 MeV achieves a speed more than ninety percent that of light and its mass becomes about 2.5 times its rest mass. It is also discussed that the relativistic effect can be overcome by varying the magnetic field, and the electron synchrotron is based on this idea. But, before that instrument is discussed, it is convenient to consider another electron accelerator, the betatron, where magnetic induction is used to accelerate the electrons.

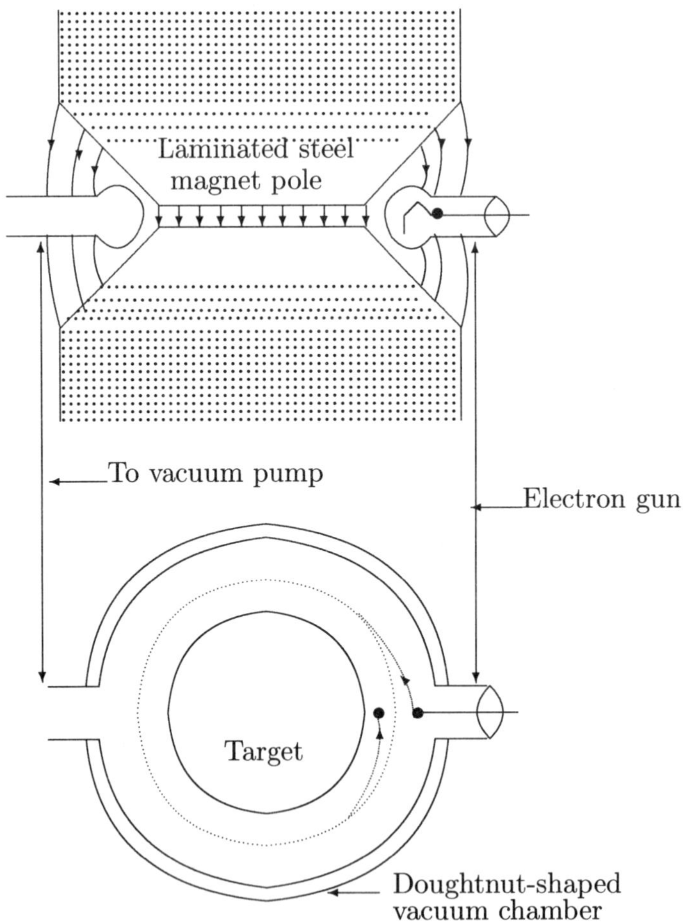

Figure 4: Accelerating chamber in a betatron.

The action of betatron is based on the same principle as that of the transformer, in which an alternating current applied to a primary coil induces a similar current in the secondary coil. Here, the primary current produces an oscillating magnetic field which, in turn, induces an oscillating potential in the secondary coil. The betatron is a transformer in which a cloud of electrons, located inside an annular, doughnut-shaped vacuum chamber behave as the secondary coil. The chamber is placed within the poles of electromagnet energized by alternating pulsed current, and the magnet produces a strong field in the central space (hole) of the doughnut. The electrons move in a circular orbit of constant radius within the vacuum chamber, as shown in Figure 4. They gain energy by induction, because of the change of magnetic flux Φ, linking the orbit, with time.

For application of this method, the induced accelerating field must have just the proper strength at the stable orbit to ensure that the electrons remain in that orbit during the increase of magnetic field. That is, the flux linking that orbit must have the right magnitude. The required relationship between the linking flux and the magnetic field can be derived in simple manner. The induced voltage per turn is $\mathrm{d}\Phi/\mathrm{d}t$ is similar to the case of transformer and the electric field and is expressed as

$$E = \frac{1}{2\pi R}\,\frac{\mathrm{d}\Phi}{\mathrm{d}t}$$

where R is the radius of the stable orbit. The force on the electron is eE, and the law of motion for the electron is

$$\frac{\mathrm{d}}{\mathrm{d}t}(mv) = eE = \frac{e}{2\pi R}\,\frac{\mathrm{d}\Phi}{\mathrm{d}t} \tag{8.7}$$

To maintain motion in a circular orbit of constant radius R, the magnetic field B at the orbit must increase as the electron momentum increases. The momentum of the electron in the field, we have

$$Bev = \frac{mv^2}{R} \qquad \text{or} \qquad mv = eRB$$

so that

$$\frac{\mathrm{d}}{\mathrm{d}t}(mv) = eR\,\frac{\mathrm{d}B}{\mathrm{d}t} \tag{8.8}$$

Acceleration by magnetic induction at constant radius can be obtained when the time rate of change of momentum is the same as that given by the time rate of change in magnetic field. In other words, the expressions on the right sides of equations (8.7) and (8.8) must be equal. Thus, we have

$$\frac{\mathrm{d}\Phi}{\mathrm{d}t} = 2\pi R^2\,\frac{\mathrm{d}B}{\mathrm{d}t} = 2\,\frac{\mathrm{d}}{\mathrm{d}t}(\pi R^2 B) \tag{8.9}$$

Equation (8.9) is the betatron condition, which states that in any time interval the linking flux Φ must changes at a rate which is twice that which would occur when the central magnetic field has been uniform and equal to the field at the orbit. It is important to note that this condition holds for both the relativistic as well as non-relativistic energies, because the law of force as written in equations (8.7) and (8.8) satisfies the requirements of special relativity. The betatron condition makes it necessary to have a central iron core with high flux density inside the orbit. Since the induced potential is determined by the rate of change of flux, the iron core is laminated as in the case of a transformer, and an alternating power at 60 or 180 cycles is used to produce the varying magnetic field.

Electrons are injected into the chamber with the help of a 'gun' in which thermo-electron from a hot cathode are accelerated and focused by a potential of some thousands of volts. The electrons are injected at the time when the magnetic field is just rising from its zero value in the first quarter cycle, as shown in Figure 5.

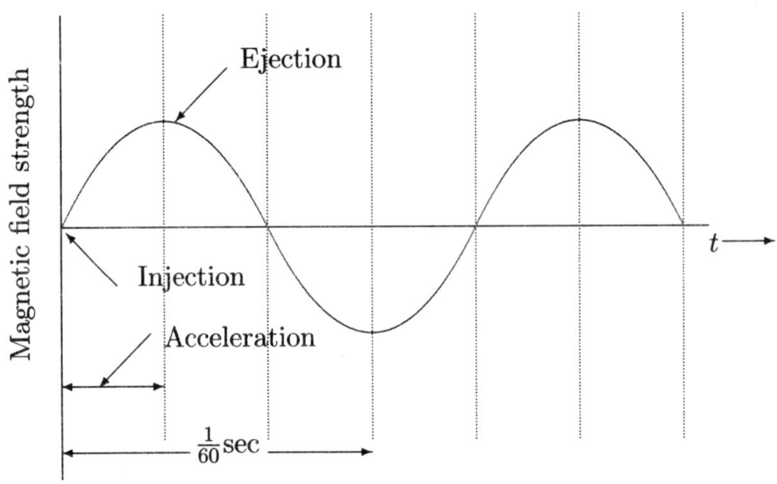

Figure 5: The acceleration cycle in a betatron.

The increasing magnetic field induces potential within the doughnut which increases the energy of electrons. Since they are traveling in a magnetic filed, the electrons move in a circular path, and the increasing magnetic field keeps them moving in a stable orbit. When the field strength passes its maximum value and starts to decrease, the direction of the induced electromagnetic force is changed and the electrons start to slow down. To avoid the phase of slowing down, the electrons are removed from their orbit when the magnetic field strength has its maximum value. A pulse of current is sent through an auxiliary coil and it changes the magnetic field; the electrons are pulled out of their stable orbit and used for the required purposes.

Exercise 5: In a betatron, the maximum magnetic field at orbit was 0.4 Wb/m^2, operating at 50 Hz with a stable orbit of diameter 1.524 m. Calculate the average energy gained per revolution and the final energy of the electrons.

Solution: In the betatron, the electron velocity is nearly c. Thus, the total distance traveled in the acceleration time (*i.e.*, one quarter cycle) is

$$= c \times \frac{T}{4} = \frac{\pi c}{2\omega}$$

Number of revolutions N are

$$N = \frac{\pi c/2\omega}{2\pi r} = \frac{c}{4\omega r}$$

Using the values

$$N = \frac{3 \times 10^8}{2\pi \times 50 \times 0.762} = 3.132 \times 10^5$$

Let E be the final energy acquired by the electrons. Since the electrons must be treated relativistically, the momentum is $mv = E/c$. Using the relation

$$Bqv = \frac{mv^2}{r} \qquad \text{we have} \qquad mv = Bqr \qquad \text{and therefore} \qquad E = Bqrc$$

Using the values

$$E = \frac{0.4 \times 1.6 \times 10^{-19} \times 0.762 \times 10^8}{1.6 \times 10{-}16} = 91.45 \times 10^6 eV$$

The average energy gained per revolution is

$$= \frac{91.45 \times 10^6}{3.132 \times 10^5} = 291.9 \text{ eV}$$

Electron synchrotron

When the electron energy increases, the electrons can be held in a stable orbit by increasing the strength of the magnetic field proportional to the increase in the mass of electron. The electron motion has property of phase stability, and the theory is similar to that for positively charged particles in the frequency modulation cyclotron. As it is not necessary to satisfy the betatron condition, we can use a ring shaped magnet. The instrument based on these ideas is known as the electron synchrotron and is shown in figure 6.

Electrons can be injected at high enough speed so that the orbit radius, although not constant, does not increase greatly. Practically, electrons could be accelerated conveniently up to 1 or 2 MeV with the help of betatron using the flux bars located inside the orbit. These bars are made of high permeability metal.

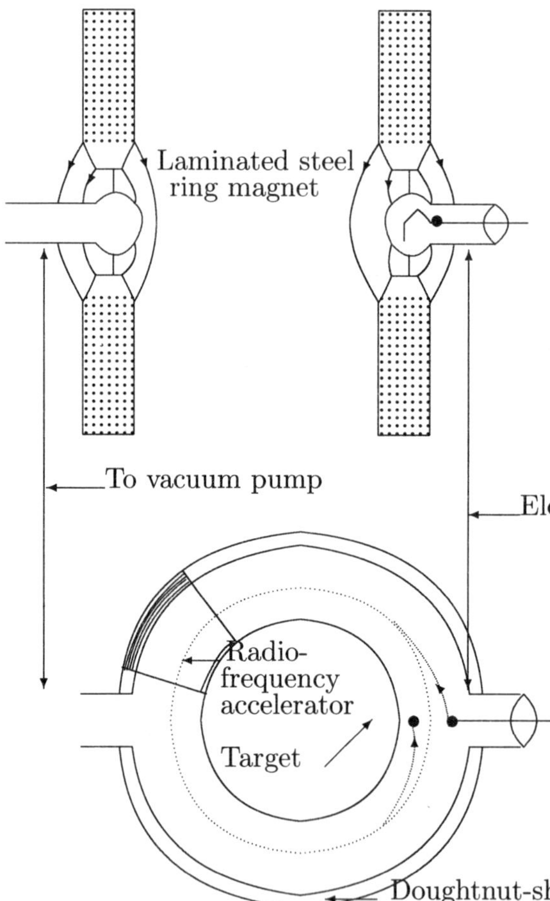

Figure 6: The accelerating chamber in a synchrotron.

Proton synchrotron

The principle and working of proton synchrotron is basically the same of the electron synchrotron. Protons revolve in an orbit of constant radius in a doughnut-shaped vacuum chamber. A ring shaped magnet produces a magnetic field normal to the chamber. Owing to large mass, protons can reach 98 present of the speed of light at about 4 BeV, and therefore, the rotation frequency for protons in an orbit of constant radius increases by a large factor during acceleration. The frequency of the applied electric field must also increase by a large factor. This is the difference between the electron synchrotron and proton synchrotron, besides the size. Protons, at low energy, are injected in a periodic pulse into the synchrotron orbit. They are accelerated by an oscillating magnetic field in resonance with the motion of ions while the magnetic field increases to its maximum.

The principle of phase stability can be utilized, and if the applied frequency is correct, the ions, on crossing the accelerating gap, oscillate in phase about an equilibrium phase which provides the proper average acceleration to keep the radius of the orbit constant. The magnet is excited periodically, and the protons are accelerated during the time when the magnetic field is increased. When they reach maximum energy, the frequency is distorted so that the orbit expands or contracts and the protons strike a target or an ejection device at the periphery.

Exercise 6: When a guide field of 1.8 Wb/m^2 is available in a proton synchrotron, calculate the radius required to attain energies of 10 GeV.

Solution: Equivalence mass of proton = 10 GeV + rest mass

$$= (10 + 0.938) \text{ GeV} = 11.75 \text{ amu} = 11.75 \times 1.67 \times 10^{-27} \text{ kg}$$

$$\text{Radius of orbit} = \frac{mv}{Bq} = \frac{11.75 \times 1.67 \times 10^{-27} \times 3 \times 10^{8}}{1.8 \times 1.6 \times 10^{-19}} = 20.31 \text{ m}$$

5. Linear accelerator

In a linear accelerator, charged particles are accelerated with the help of an oscillating electric field, applied to a series of electrodes. The alternate electrodes are connected together to two terminals which are connected to source of oscillating electric field (Figure 7). That is, the odd number cylinders are connected at one terminal whereas the even number cylinders are connected at another terminal. This figure shows the earliest form the linear accelerator having a set of cylindrical electrode of increasing length. In order to get acceleration, the applied frequency must be in the resonance with the motion of the particles. Ions obtained from a discharge tube, situated at one end, move along the axis of the tubes and are accelerated while moving in the gaps between the successive tubes. They are not accelerated inside the tubes as the potential is constant there.

Let us consider that the ions are positively charged and moving from left to right, as in Figure 7. When the first cylinder is negative, the ions are accelerated till they reach the first cylinder. Then they travel through the first cylinder at a speed which is constant but greater than the speed at which they were emitted. The first cylinder is just long enough so that when the ions reach the gap between the first and second cylinders, the potentials are reversed. The first cylinder is now negative positive and the second positive, so that the ions are again accelerated, this time in the gap between the first and second cylinders. The ions can be kept in phase with the reversals of potential by making successive cylinders longer to allow for the increasing speed of the ions, with the result that the ions gain additional energy each time they pass through the gap between two successive cylinders.

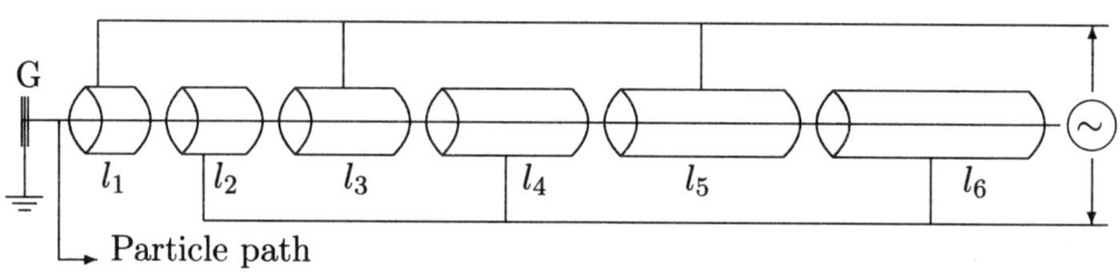

Figure 7: Schematic diagram of a linear accelerator where electrodes are in form of cylinders.

The early linear accelerators were used only for heavy, slowly moving ions such as mercury ions, because the oscillators with high frequencies were not available in those days. With the development of high frequency oscillators, linear accelerators could be designed for acceleration of protons and even electrons which can be can be accelerated to the energies up to several 100 MeV. These machines have both advantages and disadvantages in comparison to the circular machines. The size and cost of the linear machine are roughly proportional to the final energy of the particles rather than to a higher power of energy, as in the case of circular machines. Thus, for very high energies, the linear machines may be more economical. Moreover, the particles emerge from a linear accelerator in the form of a well collimated beam, automatically. On the other side, the ejection of particles in the circular machines is a bit difficult.

The main disadvantage of a linear accelerator is the fact that an individual particle, instead of passing through the same alternating field again and again, as in the circular machines, would require a large of electrodes covering very area. Here, a very long and elaborate tube need to be built. Hence, there is a difficult design problem in the adjustment of the phase of the oscillating field over a great length of accelerator, so as to ensure that the field will keep in the step with the particles. A linear accelerator can be used to obtain energies in the range of few BeV.

Exercise 7: An oscillating electric field of potential V_0 and angular frequency ω is applied between the adjacent electrodes placed in series in a linear oscillator as shown below. Derive the expressions for the lengths l_1, l_2, l_3, l_4, l_5, l_6 of the electrodes so that a particle with charge q could be accelerated.

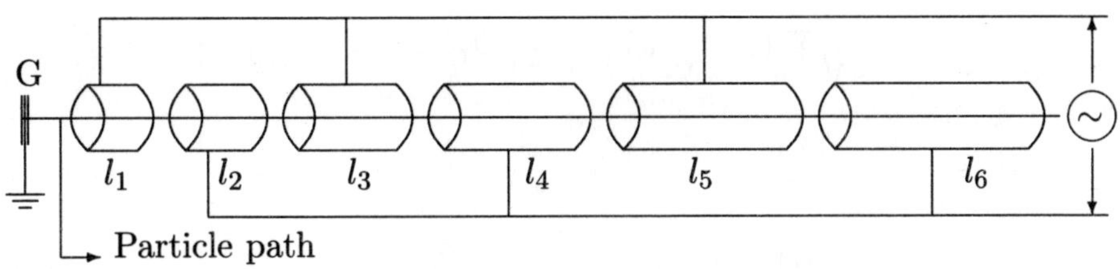

Solution: For the oscillating electric field with angular frequency ω, the time for the reversibility of the potential difference between two adjacent electrodes is

$$t = \frac{T}{2} = \frac{\pi}{\omega}$$

If particle having charge q enters an electrode with velocity v, the required length of the electrode for having resonance is

$$l = vt = \frac{v\pi}{\omega}$$

While moving through a potential difference V_0, the energy gained by the particle having charge q is qV_0. Suppose the separation between two successive electrodes is d, then we have

$$\text{energy} = \text{force} \times \text{distance}$$

$$qV_0 = mad \qquad \text{or} \qquad a = \frac{qV_0}{md}$$

Here, a and m are the acceleration and mass of the particle. If v_1 is the velocity of the particle at the exit from the previous electrode and v_2 the velocity at the entering into the following electrode, then we have

$$v_2^2 = v_1^2 + 2ad \qquad \text{or} \qquad v_2^2 = v_1^2 + \frac{2qV_0}{m}$$

It shows that the velocity of the particle does not depend on the separation between two successive electrodes. For the particle entering the electrode with velocity v_2, The required length is

$$l = \frac{v_2 \pi}{\omega}$$

Suppose, the electrons are emitted from the gun G with zero velocity. From the above discussion, the length of various electrodes are obtained as the following.

Electrode	v_1	v_2	l	ratio of length
1	0	$\sqrt{\frac{2qV_0}{m}}$	$\frac{\pi}{\omega}\sqrt{\frac{2qV_0}{m}}$	1
2	$\sqrt{\frac{2qV_0}{m}}$	$\sqrt{\frac{4qV_0}{m}}$	$\frac{\pi}{\omega}\sqrt{\frac{4qV_0}{m}}$	$\sqrt{2}$
3	$\sqrt{\frac{4qV_0}{m}}$	$\sqrt{\frac{6qV_0}{m}}$	$\frac{\pi}{\omega}\sqrt{\frac{6qV_0}{m}}$	$\sqrt{3}$
4	$\sqrt{\frac{6qV_0}{m}}$	$\sqrt{\frac{8qV_0}{m}}$	$\frac{\pi}{\omega}\sqrt{\frac{8qV_0}{m}}$	$\sqrt{4}$
5	$\sqrt{\frac{8qV_0}{m}}$	$\sqrt{\frac{10qV_0}{m}}$	$\frac{\pi}{\omega}\sqrt{\frac{10qV_0}{m}}$	$\sqrt{5}$
6	$\sqrt{\frac{10qV_0}{m}}$	$\sqrt{\frac{12qV_0}{m}}$	$\frac{\pi}{\omega}\sqrt{\frac{12qV_0}{m}}$	$\sqrt{6}$

Exercise 8: In a linear accelerator, a proton accelerated thrice by a potential of 40 kV leaves a tube and enters an accelerating space of length 30 cm before entering the next tube. Calculate the frequency of the voltage and the length of the tube entered by the proton.

Solution: Let v_1 and v_2 be the velocities of proton on entering and leaving the accelerating space. When e and m be, respectively, the charge and mass of proton. We have

$$\frac{1}{2}\,mv_1^2 = 3 \times e \times 40000 \qquad \text{or} \qquad v_1 = \Big(\frac{2.4 \times 10^5 \times e}{m}\Big)^{1/2}$$

Using the charge and mass of the particle, we get

$$v_1 = \Big(\frac{2.4 \times 10^5 \times 1.6 \times 10^{-19}}{1.67 \times 10^{-27}}\Big)^{1/2} = 4.795 \times 10^6 \text{ m/s}$$

Further, we have

$$\frac{1}{2}\,mv_2^2 = 4 \times e \times 40000$$

and thus,

$$v_2 = \left(\frac{3.2 \times 10^5 \times 1.6 \times 10^{-19}}{1.67 \times 10^{-27}}\right)^{1/2} = 5.537 \times 10^6 \text{ m/s}$$

The mean velocity while traveling 0.3 m distance $= 5.166 \times 10^6$ m/s

Time taken to travel 0.3 m is half the period ($T/2$) of the applied voltage. Thus, we have

$$\frac{T}{2} = \frac{0.3}{5.166 \times 10^6} \qquad \text{or} \qquad T = 1.161 \times 10^{-7} \text{ s}$$

Thus, the frequency of voltage applied is

$$f = \frac{1}{T} = \frac{1}{1.161 \times 10^{-7}} = 8.608 \times 10^6 \text{ Hz}$$

The protons travel through the next tube for half a period with a velocity $v_2 = 5.537 \times 10^6$ m/s. Thus, the length of the tube entered by the proton is

$$L = \frac{5.537 \times 10^6}{2 \times 8.608 \times 10^6} = 0.3216 \text{ m}$$

6. Problems and questions

1. What is an electrostatic accelerator? Describe the working of van de Graaff generator.

2. Describe the working of cyclotron. Discuss its limitation.

3. What is synchro-cyclotron? Discuss the process of phase instability.

4. Describe the working of betatron and electron synchrotron.

5. Describe the working of linear accelerator. Discuss about the advantages and disadvantages of a linear accelerator over a circular accelerator.

6. Write notes on the following

   (i) van de Graaff generator
   (ii) Cyclotron
   (iii) Betatron
   (iv) Electron synchrotron
   (v) Phase stability
   (vi) Linear accelerator

IX. Reactors

A nuclear reactor is a device in which the nuclear (atomic) energy is produced and then converted into the heat energy. The nuclear energy in a nuclear reactor is produced through nuclear chain reactions, which are initiated, controlled, and sustained at a steady rate. In this chapter, we shall discuss about some general aspects of nuclear reactors.

1. Nuclear reactor

The first nuclear reactor was constructed by E. Fermi and his coworkers. The purpose of that reactor was not only academic and research, but also to solve the problem of energy requirement. A nuclear reactor may be considered as a furnace where isotopes of uranium or plutonium are used like a fuel through the fission process. As a result of this fission, one gets neutrons, isotopes of light elements and enormous amount of energy in the form of heat. This heat can be used for conversion of water into steam, which can run the turbines used for production of electricity. In order to understand about a reactor, let us first discuss about some general aspects of a nuclear reactor.

1.1 Fission cross section

In a fission process, when a neutron is captured by a fissile nucleus (*e.g.*, U^{233}, U^{235}, U^{238} or Pu^{239}), it does not necessarily lead to fission, even when the required energy is available. It is because the compound nucleus formed, after capturing the incident neutron, may not always go through the fission process. The compound nucleus may deexcite to the ground state after releasing γ-radiation. For example, when a neutron is captured, *e.g.*, by U^{235} forming a compound nucleus U^{236}, which is in the excited state. Now, for the compound nucleus, predominantly two processes are possible: (i) radiative capture (n^1, γ) reaction, deexciting U^{236} to its ground state, and (ii) fission process, breaking the compound nucleus into two light nuclei and releasing some neutrons. Each of the two processes is characterized by its own cross-section (probability). Such cross-sections depend on the energy of neutron used.

1.2 Nuclear chain reaction

We are aware of the fact that in a fission process, the neutrons are produced. The neutrons so produced can induce fission in other fissile nuclei, thereby inducing a chain reaction. Since, on average two neutrons (more accurately 2.5) are produced per fission of a nucleus, all of them may not be available to cause further fission. Some of them may be absorbed in non-fission processes and some may escape (leak) from the system without causing a fission. However, the situation is so arranged that at least one of the neutrons per fission is available to produce another fission. This situation is known as the self supporting or a self sustained chain reaction. In other words, when the rate of production of neutrons is equal to the rate of loss of neutrons, the reaction is said to be self-supporting or self-sustained.

Nuclear reactor is a device where a self-sustained and controlled chain reaction is carried out. It was thought to use natural uranium as fuel for a nuclear reactor. Natural uranium has two isotopes, U^{238} and U^{235}, in the ratio of 99.3% and 0.7%, respectively. The difficulty with U^{238} is that it absorbs most of the neutrons without producing fission (except when a neutron has energy more than 1 MeV). On the other side, U^{235} has very large cross section for low energy neutrons. Thus, U^{235} in the natural uranium can be used in a nuclear reactor, when the neutrons produced in the fission process are slowed through collisions with some slowing down medium or moderating medium, before the neutron is able to interact with other U^{235} nucleus. Graphite is generally used as a moderator.

For a self-sustained fission chain reaction, the rate of production of neutrons must be equal to the rate of loss of neutrons. In a reactor, the loss of neutrons is either through leakage from the assembly or through absorption by the medium. The rate of leakage is proportional to the surface area whereas the rate of production is proportional to the volume of the assembly. Thus, for a self-sustained system, there should be an optimum size and optimum mass of the fissionable material. This size is known as the critical size and the mass as the critical mass.

The condition for a self-sustained fission chain reaction is expressed in terms of a parameter k, known as a neutron multiplication factor or reproduction factor and defined as

$$k = \frac{\text{Number of neutrons produced in one generation}}{\text{Number of neutrons produced in the previous generation}}$$

There are three possibilities for the value of k, as the following.

(i) When $k > 1$, the number of neutrons will increase with time and the chain reaction will become divergent and uncontrollable. Under this condition, the reactor is said to be super-critical.

(ii) When $k < 1$, the number of neutrons decreases with time and finally the chain reaction stops. Under this condition, the reactor is said to be sub-critical.

(iii) When $k = 1$, the number of neutrons in any two successive generations remains unchanged and the chain reaction continues at a steady rate. Under this condition, the reactor is said to be critical.

Because of these conditions, k is sometimes known as the critical factor.

1.3 Neutron life cycle in a reactor

A thermal neutron can cause fission of U^{235} and on average 2.5 neutrons per fission are created. Let us look into the matter what happens to these released neutrons in an assembly containing natural uranium which consists of 99.3% U^{238} and 0.7% U^{235}. Once a neutron is produced in the fission reaction, it may have the following possibilities.

(i) It may escape from the assembly (leakage).

(ii) It may be absorbed by fissile material to cause either a fission reaction or a radiative capture reaction.

(iii) It may be absorbed by non-fissile material in the assembly (like moderator, control material etc.)

Thus, in a nuclear reactor, the neutron population at any instant is a function of the rate of production of neutrons (due to fission process) and the rate of loss of neutrons (due to non-fission absorption mechanisms and leakage from the system). Now, there may be two situations: (1) the size of assembly is infinite and (2) the size of assembly is finite.

1. Infinite assembly

When the size of assembly is very large, the leakage of neutrons from its surface will be almost zero. Now, the assembly contains almost fissile material (natural uranium) and moderator (low atomic mass material like graphite). Let η be the number of fast neutrons (each with energy more than 1 MeV) produced in the fission process of U^{235} through the thermal neutron. The process of slowing down of neutrons continues. Suppose ϵ is the multiplication factor for getting the number of neutrons which slow down past the fission threshold of U^{238}. This ϵ is known as the fast fission factor. The value of ϵ is generally greater than 1, as some neutrons were already below the fission threshold of U^{238}. Now, the number of neutrons is $\eta\epsilon$, which is generally larger than η. Now, during the slowing down process, some neutrons may be captured in non-fission processes. It is known as the resonance absorption without producing new neutrons. Let p denotes the fraction escaping this absorption. Then the number of neutrons which will reach thermal energies are $\eta\epsilon p$. This factor p is also known as the resonance escape probability.

When these $\eta\epsilon p$ neutrons reach thermal energies, they will diffuse for some time in the infinite system and some of them are captured by moderator and other (non-fissionable) materials and a fraction f of these thermal neutrons is again absorbed by the fissile material. The parameter f is known as the thermal utilization factor. Thus, out of $\eta\epsilon p$ thermal neutrons, the number absorbed by the fissile material inducing further fission is $\eta\epsilon pf$. Thus, we started from one thermal neutron for producing fission at the next generation and we have $\eta\epsilon pf$ thermal neutrons capable of causing further fission. Thus, at each generation, the number of thermal electrons absorbed is $\eta\epsilon pf$. This is denoted by k_∞, defined as the infinite (medium) multiplication factor or reproduction factor for the infinite system. Thus, we have

$$k_\infty = \eta\epsilon pf$$

This relation is known as the four factor formula.

Note: In case of a reactor, the values of various quantities are $\eta = 1.31$, $\epsilon = 1.02$ $p = 0.82$ and $f = 0.88$. Then the value of k_∞ is

$$k_\infty = \eta\epsilon pf = 1.31 \times 1.02 \times 0.82 \times 0.88 = 1.05$$

(2) Finite assembly

When the size of assembly is finite, there can be leakage of neutrons, produced in the fission reaction. The leakage of fast neutrons can take place when they are slowing down and that of thermal neutrons when they diffusing in the system. Suppose, P_f and P_t denotes the non-leakage probabilities of fast and thermal neutrons, respectively. Thus, the multiplication factor for a finite system is

$$k = \eta\epsilon pfP_fP_t = k_\infty P$$

where $P = P_f P_t$ is total non-leakage probability. This relation is known as the six factor formula. Since P is always less than 1, for the critical condition, it is desirable that k_∞ must be greater than 1, such that $k_\infty P = 1$.

1.4 General features of a nuclear reactor

(a) **Fuel:** The fissile material which undergoes the fission process in a reactor is known as a fuel. The commonly used fuels are the uranium isotopes (${}_{92}U^{233}$, ${}_{92}U^{235}$ and ${}_{92}U^{238}$), thorium isotopes (${}_{92}Th^{232}$) and plutonium isotopes (${}_{94}Pu^{239}$). As compared to uranium and thorium, plutonium is considered to be much better fuel, because it gives the fission chain reaction by interacting with both slow as well as fast neutrons. The plutonium can be obtained from ${}_{92}U^{238}$ according to the following nuclear reaction.

$$ {}_{92}U^{238} + {}_{0}n^{1} \rightarrow {}_{92}U^{239} \rightarrow {}_{93}Np^{239} \rightarrow {}_{94}Pu^{239} $$

The central part of the nuclear reactor where the fission process takes place is termed as the core.

(b) **Moderator:** A moderator is a material used for slowing down the fast neutrons produced in the fission process to the thermal energies. In particular, the moderators are required in the thermal reactors. The material of a moderator should have low mass number, so that neutron could lose significant amount of energy during successive collisions with the moderator nuclei, till they attain a thermal equilibrium with the material. Further, a moderator should have a small cross section for neutron absorption and large cross section for scattering. Most commonly used moderators are graphite, heavy water (D_2O), light water (H_2O), beryllium, beryllium oxide, hydrites of materials, such as zirconium hydrite (ZrH_2) and some organic liquids, such as polyphenyls.

(c) **Control roads:** Control roads (or plates) are used to control the chain reaction and to maintain the multiplication factor, so that one could ensure a safe operation of the reactor. The control roads are generally made of materials, such as cadmium, boron, hafnium, gadolinium. They have a large cross section for neutron absorption.

The shape and form of the control roads varies from reactor to reactor. For example, in a boiling water thermal reactor, boron is used in the form of boron carbide (B_4C) powder stuffed in the stainless steel control rods having a cross section in the form of a plus (+) sign. In some reactors, Ag-In-Cd alloy is used. The control roads are moved in and out of the reactor to control the growth of neutron population.

(d) Coolant: Coolant is a material which is circulated in the reactor for removing the heat generated during the fission process in the core (as well as heat from any other parts of the reactor). A coolant can be either a liquid or a gas. A coolant in the reactor should have low cross section for neutron absorption, high thermal capacity, good thermal and radiation stability, and should not have any chemical reaction with the materials where it comes in the contact. In a thermal reactor, the coolant in general are the ordinary water, steam, heavy water, carbon-di-oxide, dry air, nitrogen, helium. In the fast reactors, the coolant in general is the liquid sodium or liquid helium.

(e) Reactor shield: In most of the reactors, neutrons and γ-rays are produced. In order to protect, the personnel and scientists working around the reactor and also equipments placed nearby, from these hazardous radiations, the reactor vessel is surrounded by a radiation shield. It is usually a thick concrete wall (about 2 m thick) capable of absorbing both neutrons and γ-rays.

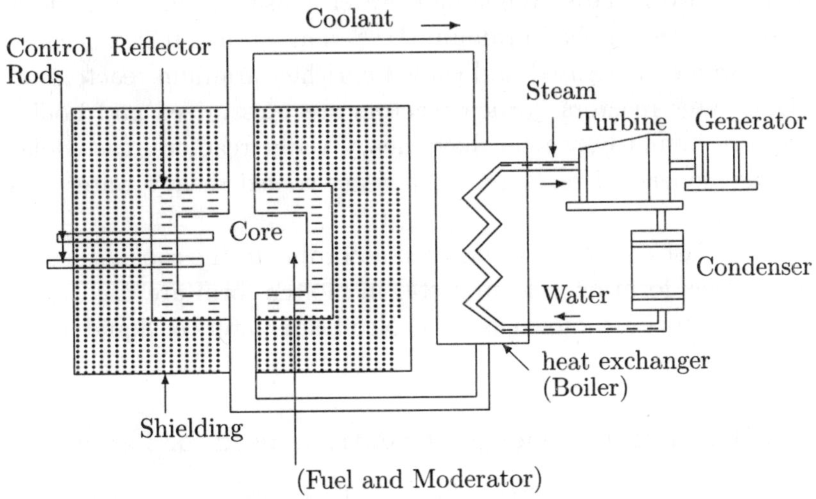

Figure 1: Schematic diagram of a nuclear reactor.

(f) Reflector: A reflector is a non-fissionable material which is generally kept near the core, in order to check the escape of neutrons from the core. In a thermal reactor, a moderating material is generally used as a reflector. In the fast reactors, Ni, Cu are used as reflectors.

A schematic diagram showing these general features of a nuclear reactor is shown in Figure 1. The heat exchanger just removes heat from the coolant, which circulates in a closed loop. The water in the heat exchanger boils and steam is generated, which is supplied to a turbine for production of electricity.

1.5 Classification of nuclear reactors

Nuclear reactors have been built for a wide range of applications, from power generation to reach and testing new components to be used in reactors. Hence, the reactors are classified in different ways. The criteria which is generally used for classification of reactors are as the following.

(i) Based on the material used for a fuel, moderator and coolant

The reactors using various combinations of fuel, moderator, coolant are being referred accordingly. For example, a thermal reactor which uses natural uranium as fuel, graphite as moderator and carbon-di-oxide gas as coolant is called as the graphite moderated gas cooled natural uranium fueled reactor. When light water is used both as a moderator and coolant with enriched uranium (U^{235} is more than 0.7%) is called as the light water moderated and cooled enriched uranium reactors or simply light water reactors. (Such reactors are pressurized and boiling water types.) Similarly, we can have heavy water reactors, gas cooled heavy water moderated reactors, light water cooled, heavy water moderated reactors.

The fast reactors are classified according to the coolant used. Helium in the gas form or sodium metal in molten form are used as coolants. Such reactors are known as gas cooled fast reactors or liquid metal cooled fast reactors.

(ii) Based on the energy of neutrons used for fission

In a nuclear reactor, according to the energy of neutron used in the fission process, a majority of nuclear reactions are classified as ‘thermal’, ‘intermediate’ and ‘fast’. In a thermal reactor, fission is induced by, so called, thermal neutrons having energy around 0.025 eV. Such reactors are generally built for research purposes and for power generation. In an intermediate (medium) energy reactor, the average energy of the neutron is more than thermal and less than few hundred electron volts (in the range 10 - 100 eV). In a fast reactor, the fission is caused by fast neutrons having energy of the order of few hundred kilo electron volts.

(iii) Based on the geometrical arrangement of fuel and moderator

According to their structure, thermal reactors are generally classified as homogeneous or heterogeneous nuclear reactors. In a homogeneous nuclear reactor, the fuel and moderator are in the same phase, both in solid or both in liquid phase. In a heterogeneous nuclear reactor, the fuel and moderator are in different phases. The fuel may be in the from of rods or plates which may be uniformly dispersed in the moderator which may be a liquid (H_2O or D_2O) or solid (graphite or BeO). Thus, the fuel and moderator are separated geometrically. Most of the present day reactors are of heterogeneous class.

(iv) Based on the main purpose of reactor

The nuclear reactors may be classified according to their utility as the following.

(a) **Basic research:** These reactors are used for research in various branches of science. Examples of such reactors are, Apsara, Cirus, Purnima, Zerlina, Dharuva. It is interesting to know that a single research reactor can serve several purposes simultaneously.

(b) **Breeder reactors:** These reactors are designed to convert fertile isotope Th^{232} or U^{238} into a fissile isotope U^{233} or Pu^{239}. They are often called the converters. When the amount of produced fissile isotope is more than that actually burnt in maintaining the fission chain reaction, the reactor is known as the breeder reactor.

(c) **Power reactors:** This class of reactors is used for production of power. These are both thermal and fast reactors.

2. Pressurized water reactor

The principle of the pressurized water reactor is based on the Clapeyron-Clausius equation for phase transition from liquid to vapour. According to this equation, when the pressure on the water surface is increased, its boiling point increases, so that it can have more quantity of heat energy per unit mass.

A block diagram of the pressurized water reactor is shown in Figure 2. The fuel uranium oxide (UO_2) is sealed in long thin zirconium-alloy tubes which are assembled together with movable control rods into a core. The core is enclosed in a steel pressure vessel. The light water is used as a coolant and moderator.

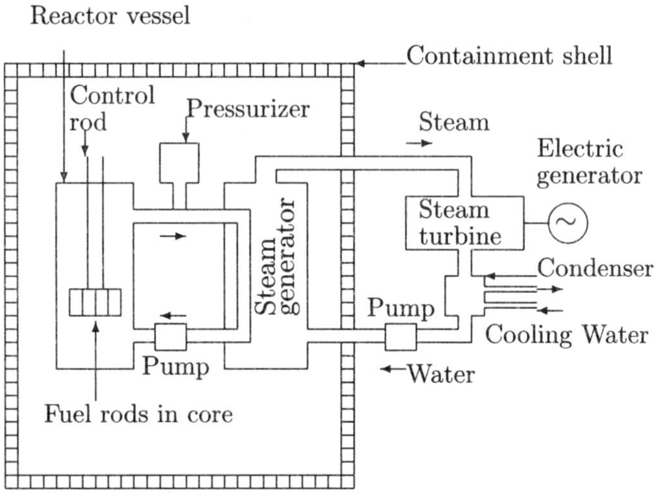

Figure 2: Block diagram of pressurized water reactor.

In order to prevent boiling, the water which circulates through the core is kept at a sufficiently high pressure of about 150 atm. The water enters the pressure vessel at about 280° C and leaves it at about 320° C. This hot water then passes through a heat exchanger (steam generator) where the steam is produced. The resulting steam then comes out of the containment shell. The containment shell serves as barrier to protect the outside world from accident to the reactor. The high pressure steam drives a turbine and the turbine drives an electric generator which produces electrical power. The steam after running the turbine gets condensed into water in the condenser. This water is recirculated through heat exchanger with the help of a pump.

3. Boiling water reactor

Uranium is used as a fuel in the boiling water reactor. A block diagram of the boiling water reactor is shown in Figure 3. The light water is used as a coolant and moderator. The circulating water absorbs heat produced by fission of the fuel and gets converted into high pressure steam in the reactor core itself. This steam runs the turbine which in turn produces electric power. The steam after running the turbine gets condensed into water in the condenser. This water is recirculated into reactor core with the help of a pump. The steam leaving the reactor may be slightly radioactive. Therefore, the turbine assembly and pipe lines are properly shielded.

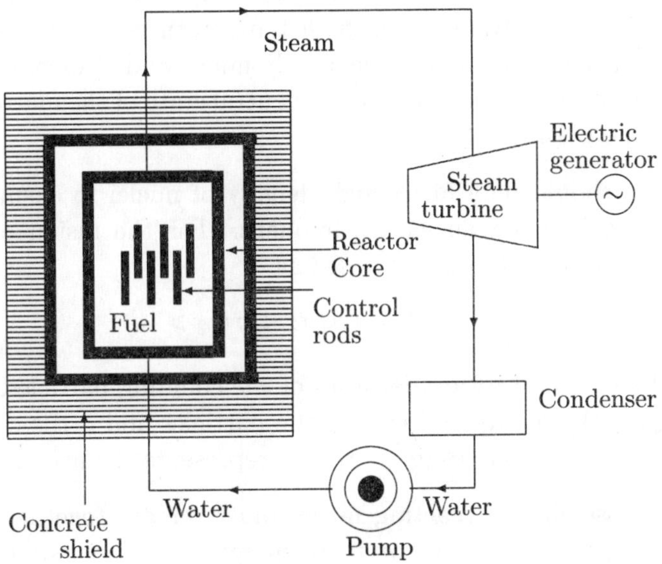

Figure 3: Block diagram of boiling water reactor.

4. Fusion reactors

Production of energy in a fusion reactor is based on the fusion reactions as the following. In the experiments for bombardment of particle ${}_1\mathrm{H}^2$ and ${}_1\mathrm{H}^3$ particles with the deuterons, the reactions are

$$ {}_1\mathrm{H}^2 + {}_1\mathrm{H}^2 \rightarrow {}_1\mathrm{H}^3 + {}_1\mathrm{H}^1 + 4.0\ \mathrm{MeV} \quad (9.1) $$

$$ {}_1\mathrm{H}^2 + {}_1\mathrm{H}^2 \rightarrow {}_2\mathrm{He}^3 + {}_0n^1 + 3.3\ \mathrm{MeV} \quad (9.2) $$

$$ {}_1\mathrm{H}^3 + {}_1\mathrm{H}^2 \rightarrow {}_2\mathrm{He}^4 + {}_0n^1 + 17.6\ \mathrm{MeV} \quad (9.3) $$

Each of the deuteron-deuteron reactions (9.1) and (9.2) has a small cross section. However, the deuteron-triton reaction (9.3) has very large cross section and also liberates a large amount of energy. The first fusion of reactors are more likely to use deuteron-triton mixture. For a successful nuclear reactor, the following conditions need to be satisfied.

1. For a fusion reaction to take place, the colliding particles should have enough energy to overcome the Colombian repulsive forces between them. It requires energies of the order of 0.01 MeV which corresponds to a temperature of 10^8 K. It shows that for a fusion reaction to take place, temperature of the order of 10^8 K should prevail initially. At such high temperatures, atoms loose almost all the electrons and we have only nuclei and electrons. This completely ionized and electrically neutral state of the matter is known as the plasma.

2. There should be a fairly high density of nuclei to ensure that collisions between nuclei are frequent. For the fusion process, the reaction rate R_{12} is expressed as

$$R_{12} = n_1 n_2 < \sigma v_{12} >$$

where n_1 and n_2 are densities of the colliding particles, say, 1 and 2, v_{12} the relative velocity of the particles and σ the cross section for the reaction. Here, $< \sigma v_{12} >$ represents the average value.

3. The assembly of reacting nuclei must remain together for a long (enough) time to pass on more energy than that required for reactor operation to take place. It is difficult to achieve high (enough) particle density n for a along (enough) time τ in a hot plasma for obtaining a net energy yield. Following the Lawson criterion, the product of the density and confinement time (*i.e.*, $n\tau$) of the plasma should be 2 $\times 10^{22}$ particle m^{-3} s and 10^{20} particle m^{-3} s for the deuteron-deuteron and deuteron-triton fusion reactions, respectively, to take place.

4.1 Advantages of fusion reactors

As a source of mechanical and electrical power, a fusion reactor has the following advantages over the fission one.

1. On the surface of the earth, the fuel (deuterium and tritium) required for a fusion reactor is much more abundant than that (uranium or thorium) required for a fission reactor.

2. Fusion reactors do not produce large amount of radioactive residues while these are the usual by products in the fusion reactors.

5. Problems and questions

1. Describe a nuclear reactor. Discuss about the general features of a nuclear reactor.

2. Write notes on the following

   (i) General features of a nuclear reactor

   (ii) Classification of nuclear reactors

   (iii) Pressurized water reactor

   (iv) Boiling water reactor

   (v) Fusion reactor

X. Elementary Particles

With the invention of high energy particle accelerators, various experiments of collision were carried out using various targets and projectiles. This led to the discovery of several new particles and new reactions. The processes and reactions involving an energy of a few MeV are generally known as the nuclear reactions (and thus belong to the nuclear physics). The reactions involving higher energy are known as the particle reactions (and thus belong to the particle physics or high energy physics).

Particle physics, also known as the 'high energy physics', is the study of elementary particles and the interactions among them. An elementary particle is one that is not made up of other particles, *i.e.,* which cannot be divided into other particles. We are aware that the matter on the earth is made up of tiny particles, called atoms and each atom is composed of a combination of three types of particles, known as the proton, neutron and electron. The proton has a positive electric charge, the neutron has no charge and the electron has a negative charge, which is equal in magnitude to that of proton. It is now well accepted that the proton and neutron are made up of even the smaller particles, called the quarks.

The forces of attraction between the particles are responsible for holding matter together. Two type of forces, the gravitational force and the electrostatic force are generally known. In addition to these two forces, there are two other forces, known as the strong force and the weak force.

According to the standard model, the true elementary particles are quarks and leptons. They interact through the strong, electromagnetic, and weak forces. The common example of leptons is the electron. Leptons are not susceptible to the strong force (they have zero value for the strong-charge[1]). The leptons include electron, muon, tau, which all have the same negative electric charge, commonly denoted by $-e$, and there is an associated neutrino for each of them. A total of six particles

[1]A particle is said to have strong-charge when it participates in the strong force. Further, a particle having weak-charge participate in the weak force.

are given in Table 1.

Table 1: Leptons and associated neutrinos

Charged lepton	Associated neutrino
electron (e)	ν_e
muon (μ)	ν_μ
Tau (τ)	ν_τ

All of the neutrinos have no electric charge, so they have zero-strong and electric charges, but non-zero weak charge.

Quarks have strong charge and are thought to be the constituents of all the observed strongly interacting particles, collectively known as the hadrons (neutrons, protons, mesons, hyperons). The strong force hold the quarks together so strongly inside of hadrons that they are unable to exist in the free form. We now know that there are six types of quarks, each given a name or flavour. These six flavours of quarks, are given the names as up (u), down (d), strange (s), charm (c), bottom/beauty (b) and top (t) in order of increasing their mass. All quarks have fractional electrical charge, $2e/3$ for u, c, and t, and $-e/3$ for d, s, and b, where e stands for the magnitude of electron charge. All quarks also carry weak charge. Each of the forces (strong and weak) has associated with it one or more carrier particles. The carriers of electromagnetic force, for example, is called the photon. The carrier of strong interaction is called the gluon - there are eight of them - and the carriers of the weak interactions are called Z^0, W^+ and W^- particles.

1. Carriers of basic forces

The elementary particles of a matter interact with one another through four distinct types of forces: gravitational force, electromagnetic force, strong force and weak force. A given particle experiences certain of these forces, while it may be immune to others. The gravitational force is experienced by all particles. The electromagnetic force is experienced only by charged particles, such as the electron and muon. The strong nuclear force is responsible for the structure of the nucleus, and only particles made up of quarks participate in the strong nuclear interaction or force. Other particles, including the electron, muon, and the three neutrinos, do not participate in the strong nuclear interactions but only in the weak nuclear interactions associated with particle decay.

Each force is carried by an elementary particle. The electromagnetic force, for example, is through the exchange of photon, the basic quantum of electromagnetic radiation. The strong force is through the exchange of gluon, the weak force through the exchange of W and Z particles, and gravitational force through the exchange of graviton. Quantum field theory applied for understanding the electromagnetic force is known as the quantum electrodynamics, and applied for the understanding of strong interaction is known as the quantum chromodynamics. In 1979 Sheldon Glashow, Steven Weinberg, and Abdus Salam were awarded the Nobel Prize in Physics for their work, where they demonstrated that the electromagnetic and weak forces are really manifestations of a single electroweak force. A unified theory that would explain that all the four forces are manifestations of a single force is being searched by the scientists.

The four forces are often naturally classified with their relative strength. Another important property of these forces is their range, *i.e.*, the distance to which the interaction is effective, and the type of quantity they couple to. The classification of these forces is given in Table 2.

Table 2: Summary of the four fundamental forces

Force	Range	Strength	Acts on
Gravitational force	∞	$\approx 6 \times 10^{-39}$	All massive particles
Electromagnetic force	∞	$1/137$	All charged particles
Strong Force	$\approx 10^{-15}$ m	≈ 1	Hadrons
Weak Force	$< 10^{-18}$ m	$\approx 1 \times 10^{-5}$	Leptons, Hadrons

2. Discovery of elementary particles

The first subatomic particle discovered is the electron. It was identified in 1897 by J. J. Thomson. After that the nucleus of an atom was discovered in 1911 by Rutherford. The nucleus of ordinary hydrogen was found to have a single particle called proton. In 1932 the neutron was discovered by Chadwick. An atom was considered to consist of a central nucleus, containing protons and neutrons, surrounded by orbiting electrons. However, other elementary particles, not found in ordinary atoms, immediately began to appear.

In 1928 the relativistic quantum theory of Dirac hypothesized the existence of a positively charged electron, positron. Mass of positron is equal to that of electron; both positron and electron have equal magnitude of charge, but opposite in nature. That is, the electron has negative charge whereas the positron has positive charge. The positron was first detected in 1932. Difficulties faced in explaining beta decay led to the prediction of neutrino in 1930, and by 1934 the existence of the neutrino was established in theory (although it was not actually detected until 1956). Another particle, called the photon, was also added to the list. It had been first suggested by Einstein in 1905 as part of his quantum theory developed for the explanation of the photoelectric effect.

The next particles discovered were related to attempts to explain the strong interactions, or strong nuclear force, binding the nucleons (protons and neutrons) together in an atomic nucleus. In 1935, Yukawa suggested that a meson (a charged particle with a mass intermediate between those of the electron and the proton) might be exchanged between the nucleons. The meson emitted by one nucleon would be absorbed by another nucleon and this would produce a strong force between the nucleons, analogous to the force produced by the exchange of photons between charged particles interacting through the electromagnetic force. It is now established that the strong force is due to the exchange of gluon. In the following year, a particle of approximately the required mass (about 200 times that of the electron) was discovered and was named the μ-meson, or muon. However, its behavior did not conform to that of the theoretical particle. In 1947, the particle predicted by Yukawa was finally discovered and named the π-meson, or pion.

Both the muon and pion were first observed in the cosmic rays. Further, studies of cosmic rays provided information about more particles. By the 1950s, these elementary particles were also observed in the laboratory as a result of particle collisions produced with the help of a particle accelerator.

The known quarks and leptons, for instance, are typically grouped in three families (where each family contains two quarks and two leptons).

These known particles are divided into two categories: baryons (heavier particles) and leptons (lighter particles) and are designated by the quantum number B or L, respectively as shown in the Table 3. Antiparticles have the same mass, rest energy, spin and isotopic spin. They have opposite charge and baryon or lepton number. They are denoted by the same symbol as the corresponding particle, but with a bar on it, except for electron (e^-) and positron (e^+).

Observations of primary cosmic rays at high altitudes and experiments with more and more powerful accelerators produced evidence for a variety of short lived particles. All of them were heavier than the pions and many of them heavier than the neutron. All of them were found not to obey some of the conservation laws, particularly the conservation of parity. They are known as the strange particles. The strange particles listed in Table 4. All the strange particles have their antiparticles. The kaon K^- is the antiparticle of K^+ and $\overline{K^0}$ is the antiparticle of K^0. The heavier particles Ξ in different charge states are called cascade particles.

Table 3: First list of elementary particles. The mass is in units of m_e, rest energy in MeV, charge in e, spin and isotopic spin (T) in units of $\hbar$; π^+ and π^- are antiparticles of each other and so are μ^+ and μ^-; π^0 is its own antiparticle.

Particle	Sy	Mass	Rest energy	Q	Spin	B or L	T
Electron	e^-	1	0.511	-1	1/2	$L=1$	0
Positron	e^+	1	0.511	$+1$	1/2	$L=-1$	0
Proton	p	1836	938.27	$+1$	1/2	$B=1$	1/2
Antiproton	$\bar{p}$	1836	938.27	-1	1/2	$B=-1$	1/2
Neutron	n	1839	939.57	0	1/2	$B=1$	1/2
Antineutron	$\bar{n}$	1839	939.57	0	1/2	$B=-1$	1/2
Pions	π^+	273.2	139.6	$+1$	0	$B=0$	1
	π^-	273.2	139.6	-1	0	$B=0$	1
	π^0	264.3	135.0	0	0	$B=0$	1
Muons	μ^+	206.8	105.66	$+1$	1/2	$L=1$	0
	μ^-	206.8	105.66	-1	1/2	$L=-1$	0
e-neutrino	ν_e	~ 0	< 17 eV	0	1/2	$L=1$	0
μ-neutrino	ν_μ	~ 0	< 250 eV	0	1/2	$L=1$	0
Antineutrinos	$\overline{\nu_e}$	~ 0	< 17 eV	0	1/2	$L=-1$	0
	$\overline{\nu_\mu}$	~ 0	< 250 eV	0	1/2	$L=-1$	0

Q denotes Charge and Sy denotes Symbol

Table 4: List of strange particles

Particle	Symbol	Rest energy	Charge	Spin	B	S	T
Heavy mesons	K^+	493.9	1	0	0	1	1/2
	K^0	497.8	0	0	0	1	1/2
Hyperons	Λ^0	1115.4	0	1/2	1	-1	0
	Σ^+	1189.4	1	1/2	1	-1	1
	Σ^0	1191.5	0	1/2	1	-1	1
	Σ^-	1196.0	-1	1/2	1	-1	1
	Ξ^0	1311	0	1/2	1	-2	1/2
	Ξ^-	1318.4	-1	1/2	1	-2	1/2

In order to understand the strange behaviour of particles, Gellmann and Nishijima proposed a strangeness scheme in which all the particles were assigned new quantum number S called the strangeness. The charge of particle is now expressed as

$$Q = T_z + Y/2$$

where the hypercharge Y is $Y = B + S$ and T_z is the component of T along the third axis. All known particles listed in Table 3 were assigned strangeness $S = 0$. The classification scheme is shown in Table 5.

Table 5: Strangeness scheme of fundamental particles

Type	S	B	Y	T	T_z	Q	Particle	Spin
Mesons	1	0	1	1/2	−1/2	0	K^0	0
	1	0	1	1/2	1/2	1	K^+	0
	0	0	0	1	−1	−1	π^-	0
	0	0	0	1	0	0	π^0	0
	0	0	0	1	1	1	π^+	0
Nucleons	0	1	1	1/2	1/2	1	p	1/2
	0	1	1	1/2	−1/2	0	n	1/2
Hyperons	−1	1	0	0	0	0	Λ^0	1/2
	−1	1	0	1	−1	−1	Σ^-	1/2
	−1	1	0	1	0	0	Σ^0	1/2
	−1	1	0	1	1	1	Σ^+	1/2
	−2	1	−1	1/2	−1/2	−1	Ξ^-	1/2
	−2	1	−1	1/2	1/2	0	Ξ^0	1/2

3. Classification of elementary particles

Two types of statistics are used to describe elementary particles, and the particles are classified on the basis of which statistics they obey. Fermi-Dirac statistics apply to those particles which obey the Pauli exclusion principle; particles obeying the Fermi-Dirac statistics are known as the fermions. Leptons and quarks are fermions. Two fermions are not allowed to occupy the same quantum state. Bose-Einstein statistics apply to all particles not covered by the exclusion principle, and such particles are known as the bosons. The number of bosons in a given quantum state is not restricted. In general, fermions compose nuclear and atomic structure, while bosons act to transmit forces between fermions; the photon, gluon, and the W and Z particles are bosons.

Basic categories of particles have also been distinguished according to other particles behaviour. The strongly interacting particles were classified as either the mesons and baryons; it is now known that the mesons consist of quark-antiquark pairs and that the baryons consists of quark triplets. The mesons class members are more massive than the leptons, but generally less massive than the protons and neutrons, although some mesons are heavier than these particles. The lightest members of the baryon class are the proton and neutron, and the heavier members are known as the hyperons. In the meson and baryon classes a number of particles that cannot be detected directly are included because their life-times are so short that they leave no tracks in a cloud chamber or bubble chamber. These particles are known as resonance, or resonance state, because of an analogy between their manner of creation and the resonance of an electrical circuit.

4. Conservation laws and symmetry

Some conservation laws apply both to elementary particles and to microscopic objects, such as the conservation of mass-energy, linear momentum, angular momentum, and charge. Other conservation laws have meaning only on the level of particle physics, including the three conservation laws for leptons, which govern members of the electron, muon, and tau families respectively, and the law governing members of the baryon class.

New quantities have been invented to explain certain aspects of particle behaviour. For example, the relatively slow decay of kaons, lambda hyperons, and some other particles led physicists to the conclusion that some conservation law prevented these particles from decaying rapidly through the strong interaction; instead they decayed through the weak interaction. This new quantity was named strangeness and is conserved in both strong and electromagnetic interactions, but not in weak interactions. Thus, the decay of a strange particle into nonstrange particles, *e.g.*, the lambda baryon into a proton and pion, can proceed only by the slow weak interaction and not by the strong interaction.

Another quantity explaining the particle behaviour is related to the fact that many particles can be described in the form of groups, called the multiplets. In a multiplet, the particles are of almost the same mass but they differ in charge. The proton and neutron form such a multiplet. The quantity describes mathematically the effect of changing a proton into a neutron, or vice versa, and is known as the isotopic spin. This name was chosen because the total number of protons and neutrons in a nucleus determines what isotope of the atom represents and because the mathematics describing this quantity is identical to that used to describe ordinary spin (the intrinsic angular momentum of elementary particles). Isotopic spin actually has nothing to do with the spin, but is represents a vector that can have various orientations in an imaginary space known as isotopic spin space. Isotopic spin is conserved only in the strong interactions.

Closely related to conservation laws, there are three symmetry principles which apply to the change of total circumstances of an event rather than the change of a particular quantity. The three symmetry operations, associated with these principles are as the following: (i) the charge conjugation (C), which is equivalent to exchanging particles and antiparticles; (ii) the parity (P), which is a kind of mirror-image symmetry involving the exchange of left and right; and (iii) the time-reversal (T), which reverses the order in which events occur. According to these symmetry principles (or invariance principles), performing one of these symmetry operations on a possible particle reaction should result in a second reaction that is also possible. However, it was found in 1956 that parity is not conserved in the weak interactions, *i.e.*, there are some possible particle decays whose mirror-image counterparts do not occur. Although one is not conserved, the combination of all three operations performed successively is conserved; this law is known as the CPT theorem.

4.1 Conservation of baryon number

The sum of baryon numbers in any process remains conserved. All normal baryons such as p^+, n^0, Λ^0, Σ^+, Σ^-, Σ^0, Ξ^-, Ξ^0, and Ω^- have a baryon number of $+1$, the corresponding anti-particles known as anti-baryons have a baryon number of -1. All the mesons have a baryon number of zero. For example, when a neutron is left to itself, it decays as the following.

$$\text{Reaction:} \quad n^0 = p^+ + e^- + \overline{\nu_e}$$

$$\text{Baryon No.} \quad +1 = +1 + 0 + 0$$

Below the reaction, we have estimated baryon numbers. The neutron and the proton are both baryons, each having a baryon number $+1$, while electron and antineutron have baryon number 0, resulting in conservation of baryon number.

4.2 Conservation of lepton number

The sum of leptons numbers in any process always remains conserved. The ordinary electron e^-, negative muon μ^- and neutrinos ν all have a lepton number $+1$, the corresponding anti-particles μ^+, e^+ and $\overline{\nu}$ known as anti leptons have a lepton number of -1. Let us consider the following reaction.

$$\begin{array}{lccc} \text{Reaction:} & \mu^+ & = & e^+ + \nu + \overline{\nu} \\ \text{Lepton No.} & -1 & = & -1 + 1 - 1 \end{array}$$

is allowed because the lepton number is conserved.

$$\begin{array}{lccc} \text{Reaction:} & n^0 & = & p^+ + e^- + \overline{\nu} \\ \text{Lepton No.} & 0 & = & +0 + 1 - 1 \end{array}$$

Below the reaction, we have estimated lepton numbers. This reaction is allowed according to the conservation of lepton numbers. Conservation of leptons has a significance for strong interactions.

4.3 Conservation of isospin

As the strong force between protons and neutrons is charge-independent, they are considered as the two states of one particle, called nucleon. This was achieved by Heisenberg by introducing a new quantum number, the isospin or isotropic spin. In the case of nucleons, $T = 1/2$ and the $2T + 1 = 2$, possible values T_z are $+1/2$ (for proton) and $-1/2$ (for neutron). For the pions, $T = 1$ and so there are $2T + 1 = 3$ charge states. The triplet consists of π^+, π^0 and π^- particles and the values of T_z are $+1$, 0 and -1, respectively.

Analytically, if B is the baryon number, T is the isotopic spin quantum number, T_z the components of T and Q the charge in units of the electron charge, the relation between these quantities is

$$Q = T_z + \frac{B}{2}$$

(Note that here the strangeness number is zero.) The isospin component T_z is conserved in both strong and electromagnetic interactions but not in weak interaction.

4.4 Conservation of strangeness number

This law states that for a reaction, the strangeness number remains unchanged in strong interaction and associated production of strange particles. Let us consider the following reaction:

$$\begin{array}{lccc} \text{Reaction:} & \pi^+ + p & = & \Sigma^+ + \mathrm{K}^+ \\ \text{Baryon No.:} & 0 + 1 & = & +1 + 0 \\ \text{Strangeness No.:} & 0 + 0 & = & -1 + 1 \end{array}$$

Below this reaction, we have estimated baryon numbers and strangeness numbers, separately. The strangeness number of π^+ and p is 0, while those of Σ^+ and K^+ are -1 and $+1$, respectively. Thus, we see that the strangeness number is conserved in the subatomic interaction. This law holds for strong and electromagnetic interactions, but not for the weak interaction.

While the conservation laws hold good for strange particles also, their reactions obey the following selection rules:

(i) $\Delta S = 0$ for strong reactions, which are fast. For example, decay reactions

$$\pi^- + p \rightarrow K^0 + \Lambda^0, \quad K^- + p \rightarrow \Xi^0 + K^0, \quad K^- + p \rightarrow \Lambda^0 + \pi^0$$

(ii) $\Delta S = \pm 1$ for weak reactions, which are slow. For example, decay reactions

$$K^+ \rightarrow \pi^+ + \pi^-, \quad \Lambda^0 \rightarrow p + \pi^-, \quad \Sigma^+ \rightarrow p + \pi^0, \quad \Xi^0 \rightarrow \Lambda^0 + \pi^0$$

4.5 Validity of conservation laws in elementary particle interactions

Having knowledge of various conservation laws, let us look about their validity in elementary particle interactions. This information has been summarized in Table 6.

Table 6: Validity of conservation laws

Conservation law	Interaction		
	Strong	Electro-magnetic	Weak
Mass-energy	Yes	Yes	Yes
Momentum	Yes	Yes	Yes
Angular momentum	Yes	Yes	Yes
Charge	Yes	Yes	Yes
Baryon number	Yes	Yes	Yes
Lepton number	Yes	Yes	Yes
Isospin T	Yes	No	No ($\Delta T = 1$ or $1/2$)
Strangeness S	Yes	Yes	No ($\Delta S = 1$ or 0)
Hypercharge Y	Yes	Yes	No
Parity P	Yes	Yes	No

Exercise 1: Which of the following reactions are allowed and forbidden under the conservation of charge, baryon number, strangeness and the

components of isotopic spin.

(1) $\pi^+ + n \rightarrow \Lambda^0 + K^+$

(2) $\pi^+ + n \rightarrow K^0 + K^+$

(3) $\pi^+ + n \rightarrow \overline{K^0} + \Sigma^+$

(4) $\pi^+ + n \rightarrow \pi^- + p$

(5) $\pi^- + p \rightarrow \Lambda^0 + K^0$

(6) $\pi^- + p \rightarrow \pi^0 + \Lambda^0$

(7) $p + p \rightarrow p + p + \pi^0$

(8) $\Lambda^0 \rightarrow \pi^- + p^+$

(9) $\pi^+ + p \rightarrow \Sigma^+ + K^+$

(10) $p + p \rightarrow \Lambda^0 + K^+ + p$

(11) $\pi^- + p \rightarrow \Sigma^- + K^+$

(12) $\pi^+ + n \rightarrow \Xi^- + K^+ + K^+$

(13) $\pi^- + p \rightarrow \pi^0 + n$

(14) $p + \gamma \rightarrow p + \pi^0$

(15) $\Sigma^0 \rightarrow \Lambda^0 + \gamma$

(16) $K^0 \rightarrow \pi^+ + \pi^-$

(17) $\pi^- + p \rightarrow \Sigma^+ + K^-$

(18) $K^- + p \rightarrow \Omega^- + K^+ + K^0$

(19) $\Omega^- \rightarrow \Xi^0 + \pi^-$

(20) $\pi^+ + p \rightarrow p + p + \overline{n}$

(21) $\Lambda^0 \rightarrow n + \gamma$

(22) $p + p \rightarrow K^+ + \Sigma^+$

(23) $p + n \rightarrow \Lambda^0 + \Sigma^+$

(24) $p + n \rightarrow \Xi^- + K^+ + \Sigma^+$

(25) $\Xi^0 \rightarrow \Sigma^0 + \Lambda^0$

(26) $\Sigma^+ \rightarrow \Lambda^0 + K^+$

(27) $\Lambda^0 \rightarrow K^+ + K^-$

(28) $p \rightarrow e^+ + \pi^0$

(29) $\Omega^- \rightarrow \Xi^0 + K^-$

(30) $\Xi^- \rightarrow \Lambda^0 + \pi^-$

(31) $K^+ \rightarrow \pi^+ + \pi^-$

Solution: For the reactions, conservation of each of the charge (Q), baryon number (B), strangeness number (S) and component of isospin (T_z) are considered in separate lines.

(1) $\pi^+ + n \rightarrow \Lambda^0 + K^+$

	π^+	+	n	Λ^0	+	K^+	
Q	1	+	0	0	+	1	$\Delta Q = 0$
B	0	+	1	1	+	0	$\Delta B = 0$
S	0	+	0	−1	+	1	$\Delta S = 0$
T_z	1	−	1/2	0	+	1/2	$\Delta T_z = 0$

Hence, reaction is allowed under the conservation laws.

(2) $\pi^+ + n \rightarrow K^0 + K^+$

Q	1	+	0	0	+	1	$\Delta Q = 0$
B	0	+	1	0	+	0	$\Delta B \neq 0$
S	0	+	0	1	+	1	$\Delta S \neq 0$
T_z	1	−	1/2	−1/2	+	1/2	$\Delta T_z \neq 0$

Hence, reaction is not allowed under the conservation laws.

(3) $\pi^+ + n \rightarrow \overline{K^0} + \Sigma^+$

Q	1	+	0	0	+	1	$\Delta Q = 0$
B	0	+	1	0	+	1	$\Delta B = 0$
S	0	+	0	−1	−	1	$\Delta S \neq 0$
T_z	1	−	1/2	−1	−	1	$\Delta T_z \neq 0$

Hence, reaction is not allowed under strangeness and the component of isotopic spin conservation laws.

(4) $\pi^+ + n \rightarrow \pi^- + p$

Q	1	+	0	−1	+	1	$\Delta Q \neq 0$
B	0	+	1	0	+	1	$\Delta B = 0$
S	0	+	0	0	+	0	$\Delta S = 0$
T_z	1	−	1/2	−1	+	1/2	$\Delta T_z \neq 0$

Hence, reaction is not allowed under charge and the component of isotopic spin conservation laws.

(5) $\pi^- + p \rightarrow \Lambda^0 + K^0$

Q	−1	+	1	0	+	0	$\Delta Q = 0$
B	0	+	1	1	+	0	$\Delta B = 0$
S	0	+	0	−1	+	1	$\Delta S = 0$
T_z	−1	+	1/2	0	−	1/2	$\Delta T_z = 0$

Hence, reaction is allowed under the conservation laws.

(6) $\pi^- + p \rightarrow \pi^0 + \Lambda^0$

Q	−1	+	1	0	+	0	$\Delta Q = 0$
B	0	+	1	0	+	1	$\Delta B = 0$
S	0	+	0	0	−	1	$\Delta S \neq 0$
T_z	−1	+	1/2	0	+	0	$\Delta T_z \neq 0$

Hence, reaction is not allowed under strangeness and the component of isotopic spin conservation laws.

(7) $p + p \rightarrow p + p + \pi^0$

Q	1	+	1	1	+	1	+	0	$\Delta Q = 0$
B	1	+	1	1	+	1	+	0	$\Delta B = 0$
S	0	+	0	0	+	0	+	0	$\Delta S = 0$
T_z	1/2	+	1/2	1/2	+	1/2	+	0	$\Delta T_z = 0$

Hence, reaction is allowed under the conservation laws.

(8) $\Lambda^0 \rightarrow \pi^- + p^+$

Q	0	−1	+	1	$\Delta Q = 0$
B	1	0	+	1	$\Delta B = 0$
S	−1	0	+	0	$\Delta S \neq 0$
T_z	0	−1	+	1/2	$\Delta T_z \neq 0$

Hence, reaction is not allowed under strangeness and the component of isotopic spin conservation laws.

(9) $\pi^+ + p \rightarrow \Sigma^+ + K^+$

Q	1	+	1	1	+	1	$\Delta Q = 0$
B	0	+	1	1	+	0	$\Delta B = 0$
S	0	+	0	−1	+	1	$\Delta S = 0$
T_z	1	+	1/2	1	+	1/2	$\Delta T_z = 0$

Hence, reaction is allowed under the conservation laws.

(10) $p + p \rightarrow \Lambda^0 + K^+ + p$

Q	1	+	1	0	+	1	+	1	$\Delta Q = 0$
B	1	+	1	1	+	0	+	1	$\Delta B = 0$
S	0	+	0	−1	+	1	+	0	$\Delta S = 0$
T_z	1/2	+	1/2	0	+	1/2	+	1/2	$\Delta T_z = 0$

Hence, reaction is allowed under the conservation laws.

(11) $\pi^- + p \rightarrow \Sigma^- + K^+$

Q	−1	+	1	−1	+	1	$\Delta Q = 0$
B	0	+	1	1	+	0	$\Delta B = 0$
S	0	+	0	−1	+	1	$\Delta S = 0$
T_z	−1	+	1/2	−1	+	1/2	$\Delta T_z = 0$

Hence, reaction is allowed under the conservation laws.

$$(12) \qquad \pi^+ + n \rightarrow \Xi^- + K^+ + K^+$$

$$\begin{array}{llllllllllll}
Q & 1 & + & 0 & & -1 & + & 1 & + & 1 & & \Delta Q = 0 \\
B & 0 & + & 1 & & 1 & + & 0 & + & 0 & & \Delta B = 0 \\
S & 0 & + & 0 & & -2 & + & 1 & + & 1 & & \Delta S = 0 \\
T_z & 1 & - & 1/2 & & -1/2 & + & 1/2 & + & 1/2 & & \Delta T_z = 0
\end{array}$$

Hence, reaction is allowed under the conservation laws.

$$(13) \qquad \pi^- + p \rightarrow \pi^0 + n$$

$$\begin{array}{lllllllll}
Q & -1 & + & 1 & & 0 & + & 0 & \Delta Q = 0 \\
B & 0 & + & 1 & & 0 & + & 1 & \Delta B = 0 \\
S & 0 & + & 0 & & 0 & + & 0 & \Delta S = 0 \\
T_z & -1 & + & 1/2 & & 0 & - & 1/2 & \Delta T_z = 0
\end{array}$$

Hence, reaction is allowed under the conservation laws.

$$(14) \qquad p + \gamma \rightarrow p + \pi^0$$

$$\begin{array}{lllllllll}
Q & 1 & + & 0 & & 1 & + & 0 & \Delta Q = 0 \\
B & 1 & + & 0 & & 1 & + & 0 & \Delta B = 0 \\
S & 0 & + & 0 & & 0 & + & 0 & \Delta S = 0 \\
T_z & 1/2 & + & 0 & & 1/2 & + & 0 & \Delta T_z = 0
\end{array}$$

Hence, reaction is allowed under the conservation laws.

$$(15) \qquad \Sigma^0 \rightarrow \Lambda^0 + \gamma$$

$$\begin{array}{llllll}
Q & 0 & 0 & + & 0 & \Delta Q = 0 \\
B & 1 & 1 & + & 0 & \Delta B = 0 \\
S & -1 & -1 & + & 0 & \Delta S = 0 \\
T_z & 0 & 0 & + & 0 & \Delta T_z = 0
\end{array}$$

Hence, reaction is allowed under the conservation laws.

$$(16) \qquad K^0 \rightarrow \pi^+ + \pi^-$$

$$\begin{array}{llllll}
Q & 0 & 1 & - & 1 & \Delta Q = 0 \\
B & 0 & 0 & + & 0 & \Delta B = 0 \\
S & 1 & 0 & + & 0 & \Delta S \neq 0 \\
T_z & -1/2 & 1 & - & 1 & \Delta T_z \neq 0
\end{array}$$

Hence, reaction is not allowed under strangeness and the component of isotopic spin conservation laws.

(17) $\pi^- + p \rightarrow \Sigma^+ + K^-$

$$\begin{array}{llllllll} Q & -1 & + & 1 & 1 & - & 1 & \Delta Q = 0 \\ B & 0 & + & 1 & 1 & + & 0 & \Delta B = 0 \\ S & 0 & + & 0 & -1 & - & 1 & \Delta S \neq 0 \\ T_z & -1 & + & 1/2 & 1 & - & 1/2 & \Delta T_z \neq 0 \end{array}$$

Hence, reaction is not allowed under strangeness and the component of isotopic spin conservation laws.

(18) $K^- + p \rightarrow \Omega^- + K^+ + K^0$

$$\begin{array}{llllllllll} Q & -1 & + & 1 & -1 & + & 1 & + & 0 & \Delta Q = 0 \\ B & 0 & + & 1 & 1 & + & 0 & + & 0 & \Delta B = 0 \\ S & -1 & + & 0 & -3 & + & 1 & + & 1 & \Delta S = 0 \\ T_z & -1/2 & + & 1/2 & 0 & + & 1/2 & - & 1/2 & \Delta T_z = 0 \end{array}$$

Hence, reaction is allowed under the conservation laws.

(19) $\Omega^- \rightarrow \Xi^0 + \pi^-$

$$\begin{array}{llllll} Q & -1 & 0 & - & 1 & \Delta Q = 0 \\ B & 1 & 1 & + & 0 & \Delta B = 0 \\ S & -3 & -2 & + & 0 & \Delta S \neq 0 \\ T_z & 0 & 1/2 & - & 1 & \Delta T_z \neq 0 \end{array}$$

Hence, reaction is not allowed under strangeness and the component of isotopic spin conservation laws.

(20) $\pi^+ + p \rightarrow p + p + \overline{n}$

$$\begin{array}{llllllllll} Q & 1 & + & 1 & 1 & + & 1 & + & 0 & \Delta Q = 0 \\ B & 0 & + & 1 & 1 & + & 1 & - & 1 & \Delta B = 0 \\ S & 0 & + & 0 & 0 & + & 0 & + & 0 & \Delta S = 0 \\ T_z & 1 & + & 1/2 & 1/2 & + & 1/2 & + & 1/2 & \Delta T_z = 0 \end{array}$$

Hence, reaction is allowed under the conservation laws.

(21) $\Lambda^0 \rightarrow n + \gamma$

$$\begin{array}{llllll} Q & 0 & 0 & + & 0 & \Delta Q = 0 \\ B & 1 & 1 & + & 0 & \Delta B = 0 \\ S & -1 & 0 & + & 0 & \Delta S \neq 0 \\ T_z & 0 & -1/2 & + & 0 & \Delta T_z \neq 0 \end{array}$$

Hence, reaction is not allowed under strangeness and the component of isotopic spin conservation laws.

(22) $p + p \rightarrow K^+ + \Sigma^+$

Q	1	+	1	1	+	1	$\Delta Q = 0$
B	1	+	1	0	+	1	$\Delta B \neq 0$
S	0	+	0	1	−	1	$\Delta S = 0$
T_z	1/2	+	1/2	1/2	+	1	$\Delta T_z \neq 0$

Hence, reaction is not allowed under baryon and the component of isotopic spin conservation laws.

(23) $p + n \rightarrow \Lambda^0 + \Sigma^+$

Q	1	+	0	0	+	1	$\Delta Q = 0$
B	1	+	1	1	+	1	$\Delta B = 0$
S	0	+	0	−1	−	1	$\Delta S \neq 0$
T_z	1/2	−	1/2	0	+	1	$\Delta T_z \neq 0$

Hence, reaction is not allowed under strangeness and the component of isotopic spin conservation laws.

(24) $p + n \rightarrow \Xi^- + K^+ + \Sigma^+$

Q	1	+	0	−1	+	1	+	1	$\Delta Q = 0$
B	1	+	1	1	+	0	+	1	$\Delta B = 0$
S	0	+	0	−2	+	1	−	1	$\Delta S \neq 0$
T_z	1/2	−	1/2	−1/2	+	1/2	+	1	$\Delta T_z \neq 0$

Hence, reaction is not allowed under strangeness and the component of isotopic spin conservation laws.

(25) $\Xi^0 \rightarrow \Sigma^0 + \Lambda^0$

Q	0	0	+	0	$\Delta Q = 0$
B	1	1	+	1	$\Delta B \neq 0$
S	−2	−1	−	1	$\Delta S = 0$
T_z	1/2	0	+	0	$\Delta T_z \neq 0$

Hence, reaction is not allowed under baryon and the component of isotopic spin conservation laws.

(26) $\Sigma^+ \rightarrow \Lambda^0 + K^+$

Q	1	0	+	1	$\Delta Q = 0$
B	1	1	+	0	$\Delta B = 0$
S	−1	−1	+	1	$\Delta S \neq 0$
T_z	1	0	+	1/2	$\Delta T_z \neq 0$

Hence, reaction is not allowed under strangeness and the component of isotopic spin conservation laws.

(27)	Λ^0	$\rightarrow$	K^+	$+$	K^-	
Q	0		1	$-$	1	$\Delta Q = 0$
B	1		0	$+$	0	$\Delta B \neq 0$
S	-1		1	$-$	1	$\Delta S \neq 0$
T_z	0		1/2	$-$	1/2	$\Delta T_z = 0$

Hence, reaction is not allowed under baryon and the component of isotopic spin conservation laws.

(28)	p	$\rightarrow$	e^+	$+$	π^0	
Q	1		1	$+$	0	$\Delta Q = 0$
B	1		0	$+$	0	$\Delta B \neq 0$
S	0		0	$+$	0	$\Delta S = 0$
T_z	1/2		0	$+$	0	$\Delta T_z \neq 0$

Hence, reaction is not allowed under baryon and the component of isotopic spin conservation laws.

(29)	Ω^-	$\rightarrow$	Ξ^0	$+$	K^-	
Q	-1		0	$-$	1	$\Delta Q = 0$
B	1		1	$+$	0	$\Delta B = 0$
S	-3		-2	$-$	1	$\Delta S = 0$
T_z	0		1/2	$-$	1/2	$\Delta T_z = 0$

Hence, reaction is allowed under the conservation laws.

(30)	Ξ^-	$\rightarrow$	Λ^0	$+$	π^-	
Q	-1		0	$-$	1	$\Delta Q = 0$
B	1		1	$+$	0	$\Delta B = 0$
S	-2		-1	$+$	0	$\Delta S \neq 0$
T_z	$-1/2$		0	$-$	1	$\Delta T_z \neq 0$

Hence, reaction is not allowed under strangeness and the component of isotopic spin conservation laws.

(31)	K^+	$\rightarrow$	π^+	$+$	π^-	
Q	1		1	$-$	1	$\Delta Q \neq 0$
B	0		0	$+$	0	$\Delta B = 0$
S	1		0	$+$	0	$\Delta S \neq 0$
T_z	1/2		1	$-$	1	$\Delta T_z \neq 0$

Hence, reaction is not allowed under charge, strangeness and the component of isotopic spin conservation laws.

5. CP violation in neutral K-decay

When we move a body along a line, it is known as the spatial transformation. When we take reflection of each point of an object to the opposite side of a fixed (but arbitrary), it is known as the parity transformation and is represented by the symbol P. One more transformation may be to take ever single particle and turn it into its anti-particle. This is known as the charge conjugation transformation and is denoted by the symbol C.

When the universe would remain unchanged after taking transformation of its components, we say that the universe is symmetric or invariant under that transformation. In any physical model of the universe the laws are represented by equations and we can prove invariance under any given transformation by performing the transformation on the equations and seeing if the resulting equations are the same as the original ones.

When we consider a universe with no particles or interaction, the physical laws are invariant under both P and C transformation. When we introduce interactions we find that the universe is not invariant under P transformations. We also find that the universe is not invariant under C. However, the invariance is obtained (almost) if we consider not just P or C, but the combined transformation CP. It is the intriguing and maddening to find that the laws are not quite invariant under CP transformation. In other words we have CP violations.

The operation reverses all additive quantum numbers such as electric charge, hypercharge, strangeness, etc. While the P-transformation invert the coordinate system and the orientation of all objects in it: $x \to -x$, $y \to -y$, $z \to -z$. That is, a particle traveling to the right in the $+x$ direction finds itself traveling to left (still in the $+x$ direction) after a P-transformation. More significantly P reverses the relationship between the intrinsic angular momentum (spin) and direction of its velocity. When the spin is aligned with the velocity the particle is referred to as having positive helicity. If the spin is anti-parallel to the velocity direction, the particle has negative helicity. Under a P-transformation, the velocity direction is reversed but the spin direction is not (as spin is a purely internal quantum number); then a positive helicity particle changes to negative helicity and vice-verse. So under a CP-transformation, a negative helicity proton becomes a positive helicity anti-proton.

6. Particle symmetries

After the discovery of a large number of elementary particles, scientists started to look for some symmetries among them and they succeeded also by expressing the symmetries in the form of some symmetrical diagrams. The groups having 8 and 10 members are found of particular interest.

6.1 Meson O^- octet

Theory suggested that eight mesons of spin-zero can be grouped together to form a meson 0^- octet as shown in Figure 1, where T_z versus Y plot is prepared. (Here, 0^- stands for zero spin and odd parity.)

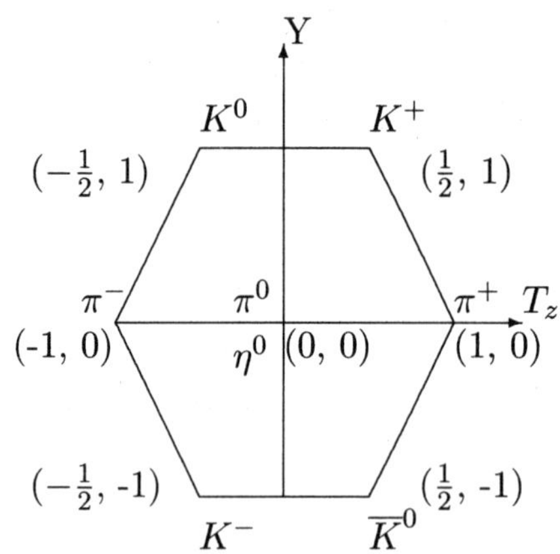

Figure 1: Meson O^- octet on T_z versus Y plot.

When it was first formulated, there were already 7 members (π^+, π^0, π^-, K^+, K^0, K^-, $\overline{K^0}$) available with a proper combination of quantum numbers. They were plotted on a T_z - Y axes to give a symmetrical hexagonal lattice (Figure 1). But, the theory required one singlet (0, 0). It was not available and that time and later on was found as η^0. The 8 members of O^- octet, along with their hypercharge Y, isospin T and component T_z, are as the following.

One triplet with	$Y = 0$	$T = 1$	$T_z = -1, 0, 1$
One doublet with	$Y = 1$	$T = 1/2$	$T_z = -1/2, 1/2$
One doublet with	$Y = -1$	$T = 1/2$	$T_z = -1/2, 1/2$
One singlet with	$Y = 0$	$T = 0$	$T_z = 0$

The members of meson O^- octet, along their quantum numbers, mass and quark constituents are given in Table 7.

Table 7: Meson O^- octet and their properties

Meson	Mass	Quantum Nos.				Quark
	MeV	S	Y	T	T_z	constituent
π^+	139.6	0	0	1	1	$u\bar{d}$
π^0	135.0	0	0	1	0	$(d\bar{d}-u\bar{u})/\sqrt{2}$
π^-	139.6	0	0	1	-1	$d\bar{u}$
K^+	493.8	1	1/2	1/2	1/2	$u\bar{s}$
K^0	497.8	1	1/2	1/2	$-1/2$	$d\bar{s}$
K^-	493.8	-1	$-1/2$	1/2	$-1/2$	$\bar{u}s$
$\overline{K^0}$	497.8	-1	$-1/2$	1/2	1/2	$\bar{d}s$
η^0	548.8	0	0	0	0	$(d\bar{d}+u\bar{u}-2s\bar{s})/\sqrt{6}$

6.2 Baryon $1/2^+$ octet

In a similar manner, the known ground state baryons with spin parity $1/2^+$ can be arranged in a hexagonal lattice as shown in Figure 2. The members of baryon $1/2^+$ octet and their properties are given in Table 8.

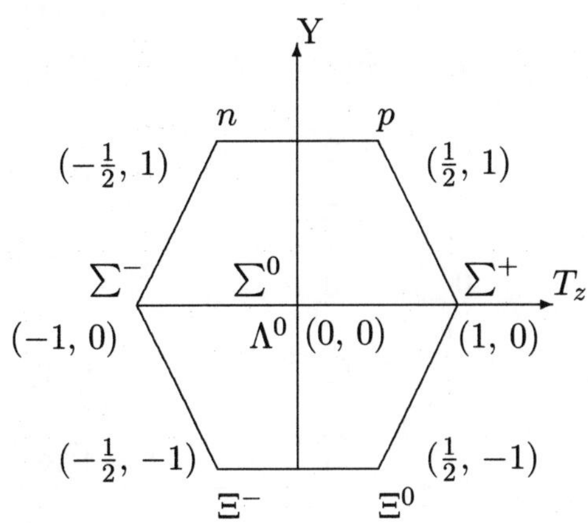

Figure 2: Baryon $1/2^+$ octet on T_z versus Y plot.

6.3 Baryon $3/2^+$ decuplet

An important prediction made on the basis of theory has been the 10-fold baryon group with spin-parity $3/2^+$. The prediction of theory is that there must be a group of 10 members with quantum numbers as the following.

One quartet with	$Y=1$	$T=3/2$	$T_z=-3/21,\ -1/2,\ 1/2,\ 3/2$
One triplet with	$Y=0$	$T=1$	$T_z=-1,\ 0,\ 1$
One doublet with	$Y=-1$	$T=1/2$	$T_z=-1/2,\ 1/2$
One singlet with	$Y=-2$	$T=0$	$T_z=0$

Table 8: Baryon $1/2^+$ octet and their properties

Meson	Mass	Quantum Nos.				Quark
	MeV	S	Y	T	T_z	constituent
p	938.3	0	1/2	1/2	1/2	uud
n	939.6	0	1/2	1/2	−1/2	udd
Λ^0	1115.5	−1	0	0	0	uds
Σ^+	1189.5	−1	0	1	1	uus
Σ^0	1192.5	−1	0	1	0	uds
Σ^-	1197.4	−1	0	1	−1	dds
Ξ^0	1314.9	−2	1	1/2	1/2	uss
Ξ^-	1321.3	−2	1	1/2	−1/2	dss

The following were the known resonances with spin-parity $3/2^+$ in 1963. They are given in Table 9 along with their quantum numbers and masses.

Table 9: Baryon $3/2^+$ decuplet and their properties

Resonance	Symbol	Q	T_z	T	Y	M (MeV)	Quark constituent
Quartet	Δ^{++}	2	3/2	3/2	1	1236	uuu
	Δ^+	1	1/2				duu
	Δ^0	0	−1/2				ddu
	Δ^-	−1	−3/2				ddd
Triplet	Σ^+	1	1	1	0	1385	uus
	Σ^0	0	0				dus
	Σ^-	−1	−1				dds
Doublet	Ξ^0	0	−1/2	1	−1	5130	uss
	Ξ^-	−1	1/2				dss
Singlet	Ω^-	−1	0	0	−2	1686	sss

Table 9 shows almost equal mass increment with the decrease of multiplicity. In the beginning, the singlet, required by the theory, was unknown. Theory predicted a particle with mass 1677 MeV, $Q = -1$, $T = 0$, $T_z = 0$, $Y = -2$, $B = 1$, $S = -3$. Later on, a particle designated as Ω^- having mass 1686 MeV was confirmed. A T_z - Y plot for baryon $3/2^+$ decuplet is shown in Figure 3.

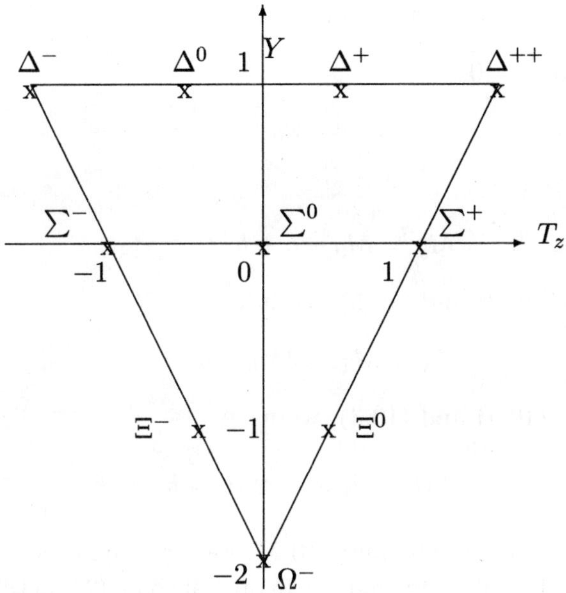

Figure 3: Baryon $3/2^+$ decuplet on T_z versus Y plot.

7. Gell-Mann-Okuba mass formula

Gell-Mann and Okuba derived a mass-formula for the several charge-multiplets observed in baryon and meson.

7.1 For baryons

The mass-formula for the baryons is as the following.

$$M = M_0 + aY + b\{T(T+1) - Y^2/4\} \tag{10.1}$$

Where M_0, a and b are constants. The spin 1/2 baryons form an octet, while spin 3/2 baryons form a decuplet.

Baryon $1/2^+$ octet

Using the formula (10.1), we can find the masses of nucleon (N), Sigma (Σ), Lambda (Λ) and Xi (Ξ) in terms of the constants and the relation between their masses.
Nucleon: $Y = 1, T = 1/2$

$$M_N = M_0 + a + b/2 \tag{10.2}$$

Sigma: $Y = 0, T = 1$

$$M_\Sigma = M_0 + 2b \tag{10.3}$$

Lambda: $Y = 0, T = 0$

$$M_\Lambda = M_0 \tag{10.4}$$

Xi: $Y = -1, T = 1/2$

$$M_\Xi = M_0 - a + b/2 \tag{10.5}$$

From equations (10.2) and (10.5), we have

$$M_N + M_\Xi = 2M_0 + b \tag{10.6}$$

From equations (10.4) and (10.3), we have

$$\frac{1}{2}\,(3M_\Lambda + M_\Sigma) = 2M_0 + b \tag{10.7}$$

Comparing equations (10.6) and (10.7), we get a relation between the masses of several multiplets that occur in Baryon $1/2^+$ octet.

$$M_N + M_\Xi = \frac{1}{2}\,(3M_\Lambda + M_\Sigma)$$

Baryon $3/2^+$ decuplet

Let us consider the various charge multiplets and their quantum numbers Y and T.

$$\begin{aligned}
\Delta &: \; Y = 1, \quad T = 3/2 \\
\Sigma &: \; Y = 0, \quad T = 1 \\
\Xi &: \; Y = -1, \quad T = 1/2 \\
\Omega &: \; Y = -2, \quad T = 0
\end{aligned}$$

Substituting the value $T = 1 + Y/2$ in equation (10.1), we get

$$M = M_0 + 2b + (a + 3b/2)Y \tag{10.8}$$

From equation (10.8), we get

$$\begin{aligned}
M_\Sigma - M_\Delta &= -(a + 3b/2) \\
M_\Xi - M_\Sigma &= -(a + 3b/2) \\
M_\Omega - M_\Xi &= -(a + 3b/2)
\end{aligned}$$

From which, it follows that

$$M_\Sigma - M_\Delta = M_\Xi - M_\Sigma = M_\Omega - M_\Xi$$

8. Quarks

A large number of elementary particles indicated that they are composed of still more fundamental particles. In 1961, deep inelastic scattering of energetic electrons by protons indicated that the proton has a structure, with three centers of fractional charge. So Gell-Mann suggested that hadrons (mesons, nucleons and hyperons) are composed of three fundamental particles, known as the quarks. They are denoted as u (up), d (down) and s (strange) quarks, which have the properties as summarized in Table 10.

Table 10: Properties of u, d, s quarks

Quark	Symbol	T	T_z	B	S	Y	Q	Spin
up	u	1/2	1/2	1/3	0	1/3	2/3	1/2
down	d	1/2	−1/2	1/3	0	1/3	−1/3	1/2
strange	s	0	0	1/3	−1	−2/3	−1/3	1/2

The antiquarks are represented by $\bar{\mathrm{u}}$, $\bar{\mathrm{d}}$, $\bar{\mathrm{s}}$, respectively. For them, the parameters T_z, B, S, Y and Q have values opposite to those for the respective quark.

- A meson is formed with a combination of one quark and one antiquark.
- A baryon is formed by a combination of three quarks.

Mesons

They form a singlet and an octet with the various quantum numbers $[T_z, B, S, Y, Q]$ shown in bracket.

Singlet:

$$(\mathrm{u}\bar{\mathrm{u}} + \mathrm{d}\bar{\mathrm{d}} + \mathrm{s}\bar{\mathrm{s}})/\sqrt{3} : [0,0,0,0,0], \quad \eta_0 - \text{singlet}$$

Octet:

$$\left.\begin{array}{lll} \mathrm{d}\bar{\mathrm{s}}: & [-1/2,0,1,1,0], & \mathrm{K}^0 \\ \mathrm{u}\bar{\mathrm{s}}: & [1/2,0,1,1,1], & \mathrm{K}^+ \end{array}\right\} \quad \text{doublet}$$

$$\left.\begin{array}{rll} \mathrm{d}\bar{\mathrm{u}}: & [-1,0,0,0,-1], & \pi^- \\ (\mathrm{u}\bar{\mathrm{u}} - \mathrm{d}\bar{\mathrm{d}})/\sqrt{2}: & [0,0,0,0,0], & \pi^0 \\ \mathrm{u}\bar{\mathrm{d}}: & [1,0,0,0,1], & \pi^+ \end{array}\right\} \quad \text{triplet}$$

$$(u\bar{u} + d\bar{d} - 2s\bar{s})/\sqrt{6} : [0, 0, 0, 0, 0], \qquad \eta_8 - \text{singlet}$$

$$\left.\begin{array}{lll} s\bar{u}: & [-1/2, 0, -1, -1, -1], & K^- \\ s\bar{d}: & [1/2, 0, -1, -1, 0], & \overline{K^0} \end{array}\right\} \quad \text{doublet}$$

One does not observe η_0 and η_8 separately, but their mixtures as η and η'.

Baryons

Since a baryon is formed by three quarks, its spin is either 1/2 or 3/2. The spin 1/2 baryons form a singlet and an octet, while spin 3/2 baryons form a decuplet, as shown below.

Singlet:

$$(\text{u d s} + \text{d s u} + \text{s u d} - \text{d u s} - \text{u s d} - \text{s d u})/\sqrt{6} : [0, 1, -1, 0, 0] \quad \Lambda' - \text{singlet}$$

Octet:

$$\left.\begin{array}{lll} \text{u d d}: & [-1/2, 1, 0, -1, 0] & \text{n} \\ \text{u u d}: & [1/2, 1, 0, 1, 1] & \text{p} \end{array}\right\} \quad \text{doublet}$$

$$\left.\begin{array}{lll} \text{d d s}: & [-1, 1, -1, 0, -1] & \Sigma^- \\ \text{u d s}: & [0, 1, -1, 0, 0] & \Sigma^0 \\ \text{u u s}: & [1, 1, -1, 0, 1] & \Sigma^+ \end{array}\right\} \quad \text{triplet}$$

$$\text{u d s}: \quad [0, 1, -1, 0, 0] \qquad \Lambda_8 \ - \text{singlet}$$

$$\left.\begin{array}{lll} \text{d s s}: & [-1/2, 1, -2, -1, -1] & \Xi^- \\ \text{u s s}: & [1/2, 1, -2, -1, 0] & \Xi^0 \end{array}\right\} \quad \text{doublet}$$

Decuplet:

$$\left.\begin{array}{lll} \text{d d d}: & [-3/2, 1, 0, 1, -1] & \Delta^- \\ \text{u d d}: & [-1/2, 1, 0, 1, 0] & \Delta^0 \\ \text{u u d}: & [1/2, 1, 0, 1, 1] & \Delta^+ \\ \text{u u u}: & [3/2, 1, 0, 1, 2] & \Delta^{++} \end{array}\right\} \quad \text{quartet}$$

$$\left.\begin{array}{lll} \text{d d s}: & [-1,1,-1,0,-1] & \Sigma^{*-} \\ \text{u d s}: & [0,1,-1,0,0] & \Sigma^{*0} \\ \text{u u s}: & [1,1,-1,0,1] & \Sigma^{*+} \end{array}\right\} \text{triplet}$$

$$\left.\begin{array}{lll} \text{d s s}: & [-1/2,1,-2,-1,-1] & \Xi^{*-} \\ \text{u s s}: & [1/2,1,-2,-1,0] & \Xi^{*0} \end{array}\right\} \text{doublet}$$

$$\text{s s s}: \quad [0,1,-3,-2,-1] \quad \Omega^- \quad - \text{singlet}$$

Some elementary particles and their quark constituent are given in Table 11.

Table 11: Particles in terms of quarks

Particle	Quarks	Particle	Quarks
Proton	uud	Ξ^0	uss
Neutron	udd	Δ^-	ddd
Σ^-	dds	Δ^0	udd
Σ^0	uds	Δ^+	uud
Σ^+	uus	Δ^{++}	uuu
Ξ^-	dss	Ω^-	sss

8.1 Need for more quarks

Around 1974 physicists discovered some new particles, such as ψ meson of 1019.41 MeV, J/ψ meson of 3096.93 MeV, γ meson of 9460.32 MeV, τ^- lepton of 1.8 GeV. Also, tau-neutrino $\nu_\tau(< 70\text{MeV})$ was suggested in 1982-83. In order to account for these particles, it was necessary to postulate the existence of three more quarks, named charm with quantum number c, top with quantum number t and beauty with quantum number b, as listed in Table 12 along with their parameters. They are, respectively, known as the charm quark, beauty quark and top quark.

Table 12: Charm, beauty and top quarks

Quark	Symbol	T	T_z	B	S	Y	c	b	Q	Spin
Charm	c	0	0	1/3	0	1/3	1	0	2/3	1/2
Top	t	0	0	1/3	0	1/3	0	1	2/3	1/2
Beauty	b	0	0	1/3	0	1/3	0	−1	−1/3	1/2

Each of them occurs in three colours, red, blue and green, and they have their antiquarks. The charge is now expressed as

$$Q = T_z + (Y + c + b)/2$$

The structure of the above particles in terms of these quarks is $\psi = (\text{s}\bar{\text{s}})$, $J/\psi = (\text{c}\bar{\text{c}})$ and $\gamma = (\text{b}\bar{\text{b}})$, but $(\text{t}\bar{\text{t}})$ has still not been found.

9. Gell-Mann matrices

Quarks have not been seen outside nuclei. They interact with and bind each other through particles called gluons which confine them within the nucleus. Gluons act through the colour property of quarks. Just as electrically charged particles interact with each other through photons according to quantum electrodynamics (QED), quarks interact with each other through gluons according to quantum chromodynamics (QCD). As there are three colours, red (R), blue (B) and green (G), expressed as

$$R = \begin{bmatrix} 1 \\ 0 \\ 0 \end{bmatrix}, \qquad B = \begin{bmatrix} 0 \\ 1 \\ 0 \end{bmatrix}, \qquad G = \begin{bmatrix} 0 \\ 0 \\ 1 \end{bmatrix}$$

and their anticolours are

$$\overline{R} = \begin{bmatrix} 1 & 0 & 0 \end{bmatrix}, \qquad \overline{B} = \begin{bmatrix} 0 & 1 & 0 \end{bmatrix}, \qquad \overline{G} = \begin{bmatrix} 0 & 0 & 1 \end{bmatrix}$$

Thus the nine possible of reactions are as the following.

$$R\overline{R} = \begin{bmatrix} 1 & 0 & 0 \\ 0 & 0 & 0 \\ 0 & 0 & 0 \end{bmatrix}, \quad B\overline{R} = \begin{bmatrix} 0 & 0 & 0 \\ 1 & 0 & 0 \\ 0 & 0 & 0 \end{bmatrix}, \quad G\overline{R} = \begin{bmatrix} 0 & 0 & 0 \\ 0 & 0 & 0 \\ 1 & 0 & 0 \end{bmatrix}$$

$$R\overline{B} = \begin{bmatrix} 0 & 1 & 0 \\ 0 & 0 & 0 \\ 0 & 0 & 0 \end{bmatrix}, \quad B\overline{B} = \begin{bmatrix} 0 & 0 & 0 \\ 0 & 1 & 0 \\ 0 & 0 & 0 \end{bmatrix}, \quad G\overline{B} = \begin{bmatrix} 0 & 0 & 0 \\ 0 & 0 & 0 \\ 0 & 1 & 0 \end{bmatrix}$$

$$R\overline{G} = \begin{bmatrix} 0 & 0 & 1 \\ 0 & 0 & 0 \\ 0 & 0 & 0 \end{bmatrix}, \quad B\overline{G} = \begin{bmatrix} 0 & 0 & 0 \\ 0 & 0 & 1 \\ 0 & 0 & 0 \end{bmatrix}, \quad G\overline{G} = \begin{bmatrix} 0 & 0 & 0 \\ 0 & 0 & 0 \\ 0 & 0 & 1 \end{bmatrix}$$

There are eight independent colour states, which correspond to the 'eight types' or 'eight colours' of gluons. There are many ways of presenting these states, which are known as the 'colour octet'. One commonly used following combinations:

$$R\overline{B} + B\overline{R}; \qquad -i(R\overline{B} - B\overline{R})$$

$$R\overline{G} + G\overline{R}; \qquad -i(R\overline{G} - G\overline{R})$$

$$B\overline{G} + G\overline{B}; \qquad -i(B\overline{G} - G\overline{B})$$

$$R\overline{R} - B\overline{B}; \qquad -i(R\overline{R} + B\overline{B} - 2G\overline{G})$$

This colour octet can be expressed in form of matrices as the following.

$$R\overline{B}+B\overline{R}=\begin{bmatrix}0&1&0\\1&0&0\\0&0&0\end{bmatrix}\qquad -i(R\overline{B}-B\overline{R})=\begin{bmatrix}0&-i&0\\i&0&0\\0&0&0\end{bmatrix}$$

$$R\overline{G}+G\overline{R}=\begin{bmatrix}0&0&1\\0&0&0\\1&0&0\end{bmatrix}\qquad -i(R\overline{G}-G\overline{R})=\begin{bmatrix}0&0&-i\\0&0&0\\i&0&0\end{bmatrix}$$

$$B\overline{G}+G\overline{B}=\begin{bmatrix}0&0&0\\0&0&1\\0&1&0\end{bmatrix}\qquad -i(B\overline{G}-G\overline{B})=\begin{bmatrix}0&0&0\\0&0&-i\\0&i&0\end{bmatrix}$$

$$R\overline{R}-B\overline{B}=\begin{bmatrix}1&0&0\\0&-1&0\\0&0&0\end{bmatrix}\qquad R\overline{R}+B\overline{B}-2G\overline{G}=\begin{bmatrix}1&0&0\\0&1&0\\0&0&-2\end{bmatrix}$$

These are equivalent to the Gell-Mann matrices. The Gell-Mann matrices are

$$\lambda_1=\begin{pmatrix}0&1&0\\1&0&0\\0&0&0\end{pmatrix}\quad \lambda_2=\begin{pmatrix}0&-i&0\\i&0&0\\0&0&0\end{pmatrix}\quad \lambda_3=\begin{pmatrix}1&0&0\\0&-1&0\\0&0&0\end{pmatrix}$$

$$\lambda_4=\begin{pmatrix}0&0&1\\0&0&0\\1&0&0\end{pmatrix}\quad \lambda_5=\begin{pmatrix}0&0&-i\\0&0&0\\i&0&0\end{pmatrix}\quad \lambda_6=\begin{pmatrix}0&0&0\\0&0&1\\0&1&0\end{pmatrix}$$

$$\lambda_7=\begin{pmatrix}0&0&0\\0&0&-i\\0&i&0\end{pmatrix}\quad \lambda_8=\begin{pmatrix}1&0&0\\0&1&0\\0&0&-2\end{pmatrix}$$

These matrices are traceless.

10. Problems and questions

1. Compare the four fundamental forces along with their range, strength and involving particles.
2. Describe Gell-Mann - Okuba mass formula for baryons and give their mass relation.
3. Describe the quarks and their properties.

4. Give detail information about the formation of mesons or baryons with the help of quark model.

5. Describe the eight colour gluons and their matrices.

6. Write notes on the following

   (i) Carriers of basic forces
   (ii) Discovery of elementary particles
   (iii) Classification of elementary particles
   (iv) Conservation of baryon or lepton number
   (v) Conservation of isospin
   (vi) Conservation of strangeness number
   (vii) CP violation
   (viii) Meson O^- octet
   (ix) Baryon $1/2^+$ octet
   (x) Baryon $3/2^+$ decuplet
   (xi) Gell-Mann-Okuba mass formulas for baryon $1/2^+$ octet
   (xii) Gell-Mann-Okuba mass formulas for baryon $3/2^+$ decuplet
   (xiii) Quarks
   (xiv) Gell-Mann Matrices

Appendices

Table 2. Some Physical Constants

	π	3.14159
Speed of light	c	2.9979×10^{8} m/s
Planck constant	h	6.62607×10^{-34} J s
Electron charge	e	1.60210×10^{-19} C
Avogadro number	N_A	6.02252×10^{26} molecules/kg mole
Boltzmann constant	k	1.38065×10^{-23} J/K
Electron rest mass	m_e	9.1091×10^{-31} kg = 511.01 keV
Proton rest mass	m_p	1.6725×10^{-27} kg = 938.26 MeV
Neutron rest mass	m_n	1.6748×10^{-27} kg = 939.55 MeV
Planck constant	h	6.6256×10^{-34} J s
Boltzmann constant	k	1.3805×10^{-23} J/K
Electron volt	1 eV	1.6021×10^{-19} J
1 eV		11,604 K = 806554.47 m^{-1}
Fermi	fm	10^{-15} m

de Broglie wavelength:

For a particle of mass m moving with velocity v, the associated wavelength is

$$\lambda = \frac{h}{mv} = \frac{h}{\sqrt{2mE}}$$

where E is the kinetic energy of the particle.

Table 3. de Broglie wavelengths

Energy MeV	Wavelength in fm		
	Photon	Electron	Proton
0.1	1.2×10^{4}	3701	90
0.5	2.5×10^{3}	1421	40
1	1.2×10^{3}	872	29
10	1.2×10^{2}	118	9
100	1.2×10^{1}	12	2.8
1000	1.2	1	0.7

Name	Symbol	Z	A	Standard Atomic weight	$f \times 10^{-4}$
Hydrogen	H	1	1	1.00794(7)	79.47
Helium	He	2	4	4.002602(2)	6.5055
Lithium	Li	3	7	6.941(2)	- 84.
Beryllium	Be	4	9	9.012182(3)	13.535889
Boron	B	5	11	10.811(7)	811.7
Carbon	C	6	12	12.0107(8)	8.983333
Nitrogen	N	7	14	14.0067(2)	4.8
Oxygen	O	8	16	15.9994(3)	-0.356250
Fluorine	F	9	19	18.9984032(5)	-0.840395
Neon	Ne	10	20	20.1797(6)	89.880000
Sodium (Natrium)	Na	11	23	22.98976928(2)	-4.448138
Magnesium	Mg	12	24	24.3050(6)	127.108333
Aluminium (Aluminum)	Al	13	27	26.9815386(8)	-6.837526
Silicon	Si	14	28	28.0855(3)	30.546429
Phosphorus	P	15	31	30.973762(2)	-8.463806
Sulfur	S	16	32	32.065(5)	20.468750
Chlorine	Cl	17	35	35.453(2)	129.485714
Argon	Ar	18	40	39.948(1)	-12.975000
Potassium (Kalium)	K	19	39	39.0983(1)	25.207692
Calcium	Ca	20	40	40.078(4)	19.60000
Scandium	Sc	21	45	44.955912(6)	-9.797200
Titanium	Ti	22	48	47.867(1)	-27.687500
Vanadium	V	23	51	50.9415(1)	-11.468627
Chromium	Cr	24	52	51.9961(6)	-0.738462
Manganese	Mn	25	55	54.938045(5)	-11.264455
Iron (Ferrum)	Fe	26	56	55.845(2)	-27.642857
Cobalt	Co	27	59	58.933195(5)	-11.322797
Nickel	Ni	28	58	58.6934(4)	119.558621
Copper (Cuprum)	Cu	29	63	63.546(3)	86.714286
Zinc	Zn	30	64	65.38(2)	215.93750
Gallium	Ga	31	69	69.723(1)	104.797101
Germanium	Ge	32	74	72.64(1)	-183.648649

Name	Symbol	Z	A	Standard	$f \times 10^{-4}$
Arsenic	As	33	75	74.92160(2)	-10.453067
Selenium	Se	34	80	78.96(3)	-129.625000
Bromine	Br	35	79	79.904(1)	114.443038
Krypton	Kr	36	84	83.798(2)	-24.023810
Rubidium	Rb	37	85	85.4678(3)	55.038824
Strontium	Sr	38	88	87.62(1)	-43.068182
Yttrium	Y	39	89	88.90585(2)	-10.578427
Zirconium	Zr	40	90	91.224(2)	136.022222
Niobium	Nb	41	93	92.90638(2)	-10.066452
Molybdenum	Mo	42	98	95.96(2)	-207.959184
Ruthenium	Ru	44	102	101.07(2)	-90.980392
Rhodium	Rh	45	103	102.90550(2)	-9.174563
Palladium	Pd	46	106	106.42(1)	39.716981
Silver (Argentum)	Ag	47	107	107.8682(2)	81.142056
Cadmium	Cd	48	114	112.411(8)	-139.315789
Indium	In	49	115	114.818(3)	-15.800000
Tin (Stannum)	Sn	50	120	118.710(7)	-107.441667
Antimony (Stibium)	Sb	51	121	121.760(1)	62.818182
Tellurium	Te	52	130	127.60(3)	-184.384615
Iodine	I	53	127	126.90447(3)	-7.521811
Xenon	Xe	54	132	131.293(6)	-53.515152
Caesium (Cesium)	Cs	55	133	132.9054519(2)	-7.108878
Barium	Ba	56	138	137.327(7)	-48.717391
Lanthanum	La	57	139	138.90547(7)	-6.800216
Cerium	Ce	58	140	140.116(1)	8.292857
Praseodymium	Pr	59	141	140.90765(2)	-6.549504
Neodymium	Nd	60	142	144.242(3)	157.908451
Samarium	Sm	62	152	150.36(2)	-107.763158
Europium	Eu	63	153	151.964(1)	-67.705882
Gadolinium	Gd	64	158	157.25(3)	-47.278481
Terbium	Tb	65	159	158.92535(2)	-4.694843
Dysprosium	Dy	66	164	162.500(1)	-91.457317
Holmium	Ho	67	165	164.93032(2)	-4.222909

Name	Symbol	Z	A	Standard	$f \times 10^{-4}$
Erbium	Er	68	166	167.259(3)	75.861446
Thulium	Tm	69	169	168.93421(2)	-3.892781
Ytterbium	Yb	70	174	173.054(5)	-54.339080
Lutetium	Lu	71	175	174.9668(1)	-1.896571
Hafnium	Hf	72	180	178.49(2)	-83.777778
Tantalum	Ta	73	181	180.94788(2)	-2.879448
Tungsten (Wolfram)	W	74	184	183.84(1)	-8.641304
Rhenium	Re	75	187	186.207(1)	-42.401070
Osmium	Os	76	192	190.23(3)	-92.031250
Iridium	Ir	77	193	192.217(3)	-40.554404
Platinum	Pt	78	195	195.084(9)	4.353846
Gold (Aurum)	Au	79	197	196.966569(4)	-1.696985
Mercury (Hydrargyrum)	Hg	80	202	200.59(2)	-69.702970
Thallium	Tl	81	205	204.3833(2)	-30.081951
Lead (Plumbum)	Pb	82	208	207.2(1)	-37.980769
Bismuth	Bi	83	209	208.98040(1)	-0.937751
Thorium	Th	90	232	232.03806(2)	1.640603
Uranium	U	92	238	238.02891(3)	1.214832

Bibliography

Abhyankar K.D. and Joshi A.W., An Overview of Basic Theoretical Physics,2009, University Press (India) Private Limited, Hyderabad

Cohen, B.L., Concepts of Nuclear Physics, 2005, Tata McGraw-Hill, New Delhi

Devanathan, V., Nuclear Physics, 2006, Narosa Publishing House, New Delhi

Ghosh, S., An Introduction to the Engineering Asoects of Nuclear Physics, 2009, I.K. International Publishing House Pvt. Ltd., New Delhi

Griffiths, D., Introduction to Elementary Particles, 1987, John Wiley & Sons, New York

Heyde, K., Basic Ideas and Concepts in Nuclear Physics, 2005, Overseas Press, India

Kaplan, I., Nuclear Physics, 1998, Narosa Publishing House, New Delhi

Kulkarni V.W., Atomic and Nuclear Physics, 2004, Himalaya Publishing House, Delhi

Tayal D.C., Nuclear Physics, 2009, Himalaya Publishing House, Delhi

Varma, J., Bhandari, R.C. and Somayajulu, D.R.S., Fundamentals of Nuclear Physics, 2005, CBS Publishers & Distributors, New Delhi

Wong, S.S.M., Introductory Nuclear Physics, 2005, Prentice-Hall, India

Index

About the authors

Professor Suresh Chandra is Professor of Physics at Lovely Professional University, Phawgara (Punjab). He obtained his M.Sc. (Physics) (1972) and Ph.D. (1977) from Meerut University (now called, Ch. Charan Singh University), Meerut. He is Fellow of the Royal Astronomical Society (London), Fellow of the Maharashtra Academy of Sciences (Pune), Associate Member of the Third World Academy of Sciences (Italy), Fellow of the Alexander von Humboldt Foundation (Germany), Member of the International Astronomical Union and Member of several other national academic bodies. He visited Germany as Alexander von Humboldt Fellow and then was in Germany as the Visiting Scientist for four and half years. Later on, he visited Germany several times under scientific exchange programme. He visited U.S.S.R. (now called, Russia) as the Guest of the U.S.S.R. Academy of Sciences. He has published 129 research papers in the journals of international repute and has written 14 books, 32 scientific articles and 8 articles pertaining to Higher Education. He has participated in 73 international/national conferences. He has completed successfully 9 research projects funded by various agencies in India and two are in progress. Seven students have been awarded Ph.D. degree under his supervision and work of three is in progress. He is member of several academic bodies in various universities. Fields of his interest as Astrophysics, Theoretical Physics and Computational Physics. He has 37 years experience of teaching and research in the universities in India as well as abroad.

Mohit Kumar Sharma is presently doing research at Department of Physics, Jiwaji University, Gwalior. He passed B.Sc. (Physics, Chemistry, Maths.) from S.R.T.M. Unkiversity, Nanded (Maharahstra) with distinction. He passed M.Sc. (Physics) from Jiwaji University, Gwalior (M.P.) with distinction. While he was M.Sc. students he was interacting with Ph.D. students and is coauthor of some research papers in the filed of astronomy & astrophysics.